Fisheries Oceanography and Ecology

Fisheries Oceanography and Ecology

Taivo Laevastu and Murray L Hayes

Northwest and Alaska Fisheries Center
National Marine Fisheries Service, NOAA
Seattle, Washington, USA

Fishing News Books Ltd
Farnham · Surrey · England

British Library CIP data
Laevastu, Taivo
Fisheries Oceanography and Ecology.
1. Fishery management, International
I. Title   II. Hayes, Murray L.
639′.22   SH328

ISBN 0 85238 117 4

Cover background picture shows some ocean currents recorded by satellite in 1981.
Courtesy of NASA.

Set by Traditional Typesetters Ltd., Chesham.
Printed in Great Britain by Page Bros (Norwich) Ltd., Norwich.
Bound by Kemp Hall Bindery Ltd., Oxford.

Contents

Figures

Tables

Preface

This book summarizes in a semi-popular manner the principal results of research in the field of fisheries oceanography during the last eight decades. It also shows how to apply facts and principles of oceanography and maritime meteorology to fisheries problems. Obviously this book is not a 'how to do' manual but rather a textbook on the subject. It is primarily meant for fisheries biologists, but it is also hoped that it will be easily understandable by and useful to skippers and administrators in the fisheries industries. The text is in general brief but is supplemented with relevant illustrations. In addition, numerous references are given to detailed descriptions.

Emphasis has been placed on the description of subjects where the interests of biologists and oceanographers must cross. These subjects are especially illustrated in the summary reviews on the influence of various environmental conditions on fish behaviour and fishery resources.

In addition, brief reviews are given of ocean analyses and forecasting and of possible fisheries diagnostic and prognostic services. Furthermore, some background of modern fisheries management on the basis of ecosystem approach is given.

Finally, it is obvious that this text cannot be an extensive treatise of all the aspects of fisheries oceanography. Moreover, the approach has been directed by the specific individual interests of the authors, but with an audacity based upon the belief that most other hypothetical authors would have been forced to approach fisheries oceanography from a narrower angle of vision.

Purpose of this book and its manner of presentation

This is the third revised edition of a book summarizing the knowledge of the relations between fish and its ocean environment published by this publisher. The two earlier editions: Hela and Laevastu, *Fisheries Hydrography*, 1962; and Laevastu and Hela, *Fisheries Oceanography*, 1970, are out of print. Unfortunately Professor Ilmo Hela has died meanwhile. Dr Murray L Hayes of the Northwest and Alaska Fisheries Center has joined with one of the earlier authors to produce this entirely revised version of the book on fisheries oceanography and marine fisheries ecology.

The fisheries of the high seas have become an internationally competitive, and economically risky business. To secure the economic survival of a given fishery it has become necessary to utilize the available technological and scientific knowledge for improving fishing conditions as well as the catch per unit effort. One of the main purposes of this book is to show ways of applying the results of research in oceanography and marine biology to benefit fisheries. It has not always been fully realized in many quarters that research is partially useless unless it is exploited. Thus, scientists should be able to use language that is understood by practical people. The specific aim of this book is to show how to apply accumulated knowledge of the environment to fisheries problems and how to utilize environmental forecasting services for fisheries.

The scientific-technical literature on the subject matter of this book has grown considerably in the twenty years since the publication of *Fisheries Hydrography* by Hela and Laevastu. This literature has also become more specific to location, fisheries, and species. It has thus become impracticable to summarize all the accumulated knowledge and experiences in detail in one book. Consequently the present authors have decided to review selectively the knowledge on the relations between fish and its environment, emphasizing the practical applications. This has been attempted by generalization and through graphical schematization, assuming that the local specialized knowledge for particular applications is available in local fisheries research institutions.

Much of the knowledge in fisheries ecology has become generally accepted. In this book these facts are presented as statements without presenting the background and proof, but giving a few examples and selected references. On the other hand, some of the fish-environment interactions are complex and may be affected simultaneously by several conditions and factors. This complexity is also pointed out where necessary.

Several chapters in this book have been taken from the last (second) edition with relatively small changes, other chapters contain major revisions or are entirely new. Earlier editions reflected heavily the conditions and research results from the North Atlantic. Emphasis remains on this area, but additional examples are added from the North Pacific.

It has become painfully apparent that the marine fisheries resources are limited and exploited near the limits of their maximum production. Thus considerable efforts are being expended on the resource conservation and on methods and means of sharing the available resources between different nations. Therefore, the final chapter of this book reviews some modern methods for fisheries resource evaluation and management.

This book covers the background of two branches of marine science – fisheries ecology and fisheries oceanography, which have considerable overlap in subject matter. Fisheries ecology is concerned with the interactions among fish, the biota, and the environment. Fisheries oceanography deals with those aspects of oceanography which can be applied to fisheries ecology, fisheries management, and practical fishing. One of the important parts of this branch of oceanography deals with the variability of oceanographic conditions in space and time. The background of the application of oceanography has been retained in this book, whereas the details of oceanographic analysis and forecasting found in previous editions have been excluded.

As the book deals with subject matter from several scientific disciplines, such as biology, meteorology, and oceanography, an explanation of terms has been included.

The scope, purpose and contents of this book will obviously not satisfy everybody concerned with fisheries, as there exists a great number of differing opinions on the interpretation of observations as well as of different ideas of the need of services to fisheries. However, the response to the earlier editions of this book has demonstrated the need for a new and somewhat enlarged edition. It is thought that this revised work might be useful to fisheries managers, fishing skippers, college students, and new scientists joining the marine fisheries research institutions.

1

Fisheries ecology – deterministic and causal or probabilistic by chance ?

Summary

The response of fish to a given environmental stimulus has usually been quantified on the bases of deterministic cause and effect. However, there exists some randomness (probabilistic results) in the reactions and responses of fish. The reasons for the varied responses are several, the main reason being that several factors (stimuli) affect the fish simultaneously and therefore we might expect deviations from an expected (normal) response to a single factor (stimulus). This variability of response must be considered in the application of fisheries oceanography to practical fisheries problems.

This book presents a summary of fisheries ecology and fisheries oceanography, mostly in a deterministic and causalistic manner. The next chapter summarizes the dependence of fish behaviour and some aspects of its physiology on selected environmental conditions, such as temperature, as cause-effect phenomena. However, the probabilistic nature of fisheries ecology is pointed out throughout this book by reporting the variability of the response (randomness) of the fish to different environmental stimuli and emphasizing that many factors might be operative simultaneously in environment-species interactions. Thus we cannot explain all the interactions on the bases of cause and effect.

Ecology has followed the path of other sciences and has attempted to study effects on the behaviour of fish of single factors in the environment (*ie* the response of fish to single stimulus). Strict experimental solutions, however, are seldom possible in the marine ecosystem, and consequently we must deduce logical dependence and must draw intuitively deterministic conclusions – *ie* postulate that the events at different times are connected with some laws which are often known only partially in such a way that some prediction is possible.

Causality requires, however, real verifiable dependence of occurrence of one event on another (*ie* dependence of a reaction to a stimulus) through repetitive observations or experiments. Even if we could eliminate all other simultaneously acting factors (stimuli)

1

but one, and could make many repetitive observations, we will find variability in our results (*ie* variability of response to stimulus) due to experimental errors. This is more so where many factors (stimuli) together affect a phenomenon (response) and we cannot eliminate these factors nor do we know the quantitative relations between these factors and the phenomenon under observations. Thus, it is imperative that in the future we make more use of probabilistic approaches in fisheries ecology (*ie* consider some degree of randomness), with the awareness that the results are open to modifications in light of new experiences.

On the other hand, we cannot assume complete randomness of action in fish as this neglects the laws of probability which are also laws of nature. There seems to be no other secure basis for recognition of the degree of deviation of a given species of fish in its response to environmental variables, except adherence to principles of objectivity in interpretation of subjective impressions and equating them with objective factors. Thus, if we recognize that causes may be interlocked in a complicated way, and objective evaluation in detail is impossible, we must partly enter the realm of metaphysics where we still strive to retain objectivity by developing logical connections between sense impressions and 'normally expected' responses.

The slow progress in ecology is partly caused by its complexity and difficulty in conducting controlled experiments, and partly because many ecological scientists are pragmatists who are content to observe a phenomenon, measure some parameters of it, and describe it in their characteristic idioms. Such interpretations reject the notion that reality can exist beyond the level of everyday experience and that objective knowledge may be obtained from subjective experiences. Thus, fishing skippers who have often a wealth of subjective experiences are often better interpreters and users of fisheries ecology than are purely objective scientists. Very few physical oceanographers who are used to a basically deterministic science have ventured into fisheries oceanographic interpretations, which involve probabilistic approaches.

There is a great variety and variation in fisheries ecology and fisheries behavioural problems are difficult to force into strict 'scientific thinking'. We can seldom make final categorical statements in fisheries oceanography and fisheries ecology. Francis (1980) pointed out that in past fisheries science more attention has been paid to developing mathematically sophisticated methods of fitting various analytical (theoretical) models than to the basic structure and assumptions of the models themselves.

However, we can reduce our knowledge of ecological phenomena to simulations (models) which are derived from empirical data as well as experiments by use of a mixture of deterministic, causal, and probabilistic approaches. The conclusion of this book in respect to application of fisheries oceanography in practical fisheries problems – *eg* in search of fish concentrations by considering the known distribution of environmental factors such as temperature and the behaviour of given species in respect to these environmental conditions – is as follows:

Although past research may suggest that a fish population reacts to an environmental stimulus in a given way, or prefers a given environment, these findings cannot be taken as quasi-absolute facts because considerable deviations from these expected reactions of the fish can occur. The reasons for such deviations can be varied, but one of the main reasons is that a multitude (at least several) of environmental factors are giving stimuli to fish concurrently. Thus, we must evaluate several concurrently occurring conditions and factors and always consider that the expected phenomenon has a given degree of probability of occurrence (*ie* is subject to 'chance').

2

Effects of environmental factors on fish

This chapter summarizes briefly some of the effects of the physical and chemical properties of the environment on fish and gives a few examples on the interactions between the fish and its environment.

Some knowledge of the effects which environmental parameters have upon the life process of fish is prerequisite to understanding how and why variability in the ocean environment influences the distribution and abundance of them. Much of our knowledge on the influence of environmental factors on the physiology of aquatic animals is the result of years of laboratory research. This research indicates that variation in certain environmental parameters exerts a profound effect on the behaviour, movement, and survival of fish and shellfish, particularly during early life of the species.

We have to base our consideration upon the unity of an organism within its environment, since the dynamics of the abundance of a species is greatly determined by this unity, and since the environment is more easily observable and predictable than the organism. However, there does not exist a stable balance between the environment and the organism, since the environmental factors are bound to vary and the adaptive properties of the organisms fluctuate. It is always important to remember that ecological approaches must be based upon physiologically correct thinking, even if we treat the results in a probabilistic (statistical) manner.

The relation between the fish and its environment can be complex indeed. First, the effects of the environment on the fish depend on the condition of the fish itself, its state of nourishment, maturation of gonads, *etc.* Furthermore, the seemingly simple causal relation of an organism to any particular component of its environment (*eg* determined in controlled laboratory conditions) is never isolated from the influence of, and interaction with, other environmental components and from some 'free will' of the fish.

In spite of the complexity of these interactions, the ecologist is frequently forced to investigate separately the effects of various environmental factors as this is often the only practical approach and is often relatively successful in spite of its limitations. Here we shall follow the same approach; we shall summarize the effects of particular environmental factors on the behaviour of fish.

4

The term 'fish behaviour' is used here in a restricted sense: referring to the reaction of shoals and, at times, of the entire stock of a given species to the prevailing environmental conditions and their changes. Of specific interest to fisheries science are the following reactions and types of group behaviour: aggregation (shoaling), dispersal, vertical migration (both diurnal and long-term), spawning and feeding migrations, passive transport by currents, degrees of activity, feeding, and spawning. We are also concerned to a lesser degree with growth, mortality, and inter- and intra-species relations.

Of practical importance for the fisheries is knowledge of the behaviour of fish in respect to those environmental factors which are easy to observe by fishermen. This knowledge of the reactions of fish to various environmental stimuli is most useful for the detection of fishable concentrations of fish and makes possible certain improvements in fishing gear and fishing methods. Therefore special emphasis has been put on these relations which can be used, directly or indirectly, for the benefit of fisheries operations and for fisheries management.

2.1 Influence of temperature on fish

Summary
Temperature is the most easily and commonly observed environmental parameter. Therefore, results of many studies are available in the literature on various fish-temperature relationships. Fish can perceive water temperature changes which are smaller than $0.1°C$. Every species has its characteristic acclimatization (optimum) temperature range and temperature tolerance limits which might change seasonally in a given stock and can be slightly different from one stock to another of the same species.

Temperature affects the rates of metabolic processes and thus modifies the activity of fish. Consequently growth and feeding rates are also affected by the temperature of the environment. Although temperature changes act as stimuli on fish, these changes may indicate also other changes in the environment, such as advective changes of water masses.

It is very unlikely that fish can orient in a horizontal temperature field. It is likely that desirable (optimum) temperature is found via vertical movements and with random and/or seasonal movements in the horizontal plane (*eg* by onshore-offshore movement). Little information is available on the possible rate of change of acclimatization temperature, *eg* in response to temperature anomalies. Nor do we understand the position of temperature influence and stimulus in the hierarchy of other stimuli – *ie* which stimulus causes strong reaction and what is the simultaneous response to several stimuli, (*eg* to temperature and food availability).

Water temperature is the most easily observed environmental factor, and therefore numerous workers have attempted to correlate the occurrence and behaviour of fish with sea water temperature and its fluctuations. Such changes in oceanic environment are, of course, often only concomitant changes of other factors, such as currents, the direct influence of which may be considerable, while the actual influence of temperature may at times be of limited significance only. Nevertheless, in most cases the temperature may serve as a most useful indicator of the prevailing and changing ecological conditions. Moreover, not only the actual temperature, with its range of fluctuations, but its horizontal and vertical gradients, varying from place to place, must be taken into account when using temperature as a direct or indirect ecological indicator. In the following a summary and an analysis are given of the influence of temperature on the behaviour and survival of fish. Information for ascertaining fish concentrations from temperature distributions will be found in other chapters. The variability of temperature in space and time is described in Chapter 8.

Fish and shellfish are cold-blooded animals and, unlike mammals and birds, their body temperature is not internally regulated but approximates to that of their surroundings. Therefore, because of the influence heat has upon chemical reactions (metabolism), environmental temperature has profound effects on the life processes of fish and shellfish, such as growth and development and swimming speed. Fish have the ability to perceive and select a limited thermal range in which they tend to congregate. This is usually the thermal range which offers them the opportunity for maximum expression of activity and is ultimately manifested in their abundance and distribution. The sense of temperature in fish seems to be well developed. Bull (1952) concluded from his careful and extensive experiments that individual teleosts perceive and react purposely to a change in water temperature of 0·03°C. Sullivan (1954) summarized the findings of various workers on the effects of temperature on the movement of fish, and on the influence of temperature on the distribution of fish, and discussed the role of the receptor mechanism of the central nervous system in the temperature response. He stated that fish select a certain temperature because of the effect of the same on their movement (activity), and concluded that the temperature change may act on fish:

(1)   as a nervous stimulus,
(2)   as a modifier of metabolic processes, and/or
(3)   as a modifier of bodily activity.

In addition it has been shown by Schmidt (1931) and various later workers that the temperature of the environment also has a definite effect on the meristic characters of fish; eg the number of vertebrae and of fin-rays increases as the temperature decreases. The effects of temperature on fish are manifested in a multitude of ways. As temperature affects bodily activity and mobility, low temperatures may affect the escape of fish from the gear. The temperature may also bring about a difference in the regional distribution of juveniles and adults, as they often have different temperature tolerance and preferences (Alverson et al, 1964). The onshore-offshore movement of demersal fish might also be triggered by temperature. Alverson *et al, op cit*, have suggested that the intrusion of colder water on to a continental shelf (upwelling) seems to cause shoreward movement of some fish, such as Dover sole and Pacific Ocean perch in the NE Pacific.

Reid (1967) has shown that sardines of genus *Engraulis* occur in waters of 6° to 22°C. Waters with this temperature cover about three eighths of the world's ocean; however, *Engraulis* inhabit less than one tenth of the area. Thus it is seen that there are other factors, besides temperature, affecting the distribution and abundance of organisms. In addition to direct influence, changes in temperature may just indicate changes in other environmental factors.

Some of the influences of temperature on fish are schematically presented in *Fig 1*. Due to extensive studies of the multitude of possible relations between the temperature of the environment and fish, references to this subject are found throughout this book.

Rates of feeding, metabolism and growth are affected not only by the availability of food but also directly by the water temperature. At suboptimal temperatures feeding activity is usually reduced. (Indirectly the temperature affects feeding by affecting the abundance of food, *eg* plankton.) It has been found that cod will not eat if the temperature is $< 1°C$, the optimum for feeding varying between 2·2 and 15·5°C. Nikolajev (1958) counted from different catches the percentage of Baltic herring which had consumed food before being caught (*Fig 2*). During the exceptionally mild winter of 1951-52 the herring continued to feed throughout the winter, which resulted in their substantial growth. During severe winters, and especially in 1953-54, the percentage of Baltic

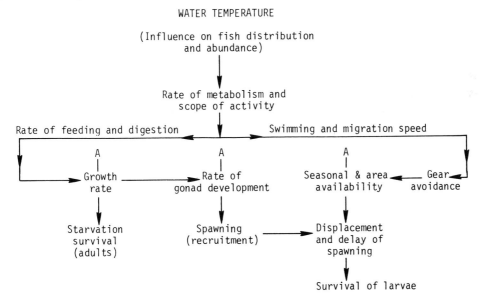

Fig 1 Schematic presentation of the effect of water temperature on the abundance, availability, and distribution of fish. Subjects marked with A are those where year-to-year anomalies occur.

7

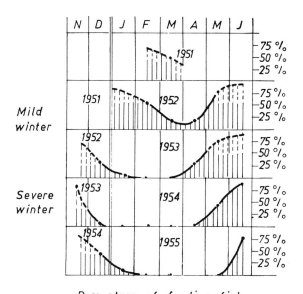

Percentage of feeding fish

Fig 2 Duration of the winter fasting period of Baltic
herring in the Gulf of Riga in the years 1952-55
(according to Nikolajev 1958). The shaded areas
indicate the number of feeding fishes. (Dashed lines
indicate extrapolation of data.)

herring which ceased to feed rose to 100% for three to four months. It has also been
observed that abnormally high temperatures cause a decrease in feeding (Komarova,
1939).

Growth is directly dependent on metabolic rates, that is, on the speed of converting
nutritive material into living matter. According to Brett (1957), there is a temperature
level at which the difference between active and resting metabolic rates is maximal.
This is the maximum activity level of fish with optimum energy release. Therefore,
there is also an optimum temperature for optimum growth of a given acclimatized
species, and the metabolic rates of this species are considerably influenced by temperature. However, the differences in the metabolic rates of acclimatized species or
populations in warm and cold seas are relatively small. Although several fish species have
a wide range of temperature tolerance, growth is usually optimal within a limited range of
higher temperatures. Kändler (1955) found that the growth of plaice is optimal at
temperatures of 13-15°C; below 1 or 2°C it is completely arrested. Johansen and Krogh
(1914) found that the rate of growth of plaice larvae is also dependent on temperature.
Figure 3 shows the experimental verification of the optimum temperature for growth of
young sockeye salmon via food conversion at different temperatures. The optimum
temperature for the growth of sockeye salmon is about 11°C.

It is frequently claimed that fish grow larger and older at low temperatures, and that
this is due to their lower metabolic rate and lower activity, or sometimes to the greater

availability of food. Taylor (1958) showed that the growth rate, life span and maximum size of cod are quantitatively related to the annual mean sea surface temperature. Wise (1959) concluded that large fish seek lower temperatures, and that this may result from a physiological need in them for lower temperatures. This theory could be applied to the interpretation of the size and age distribution of a species in the following way: the larger and older specimens migrate to the colder boundaries of the distribution area of the species due to their physiological need, while smaller specimens remain in the normal distribution area. Often the large specimens do not return to the normal spawning grounds and their spawning in the colder area is not always successful. If the above hypothesis should be proven fully true, these old stocks could be fished more intensively than 'normal' stocks. It has also to be remembered that growth is slower at lower temperatures. Thus the total biomass growth of a given species is slower at higher latitudes.

Jonsson (1965) showed that immature cod in Icelandic waters provide a good sample of a strong positive correlation between temperature and growth. The same fact was substantiated with West Greenland cod by Hermann and Hansen (1965).

The effects of temperature on the activity of fish are summarized in Chapter 5 and the practical aspects between temperature and fish distribution are described in Chapters 3 §3.3, 6 and 7.

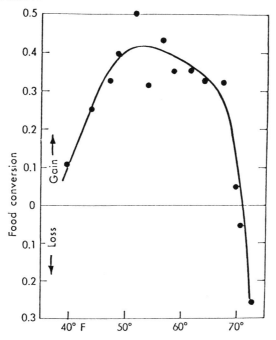

Fig 3 Growth of young sockeye salmon in two-week periods as shown by food conversion at different water temperatures (from Rounsefell 1975).

9

2.2 Effects of currents on fish

Summary

Many laboratory experiments have indicated that fish do not respond, either to direction or velocity of the flow of water, unless they have visual reference (*eg* bottom). On the other hand, field observations indicate that fish usually head into the current. It has been theorized, but not fully proven, that larger fish (*eg* tuna and salmon) can use current or geoelectric field created by current for orientation during their long migrations. It has been lately demonstrated, using sonic tags, that demersal fish can also use tidal currents for their migrations.

Currents transport eggs, larvae, and smaller fish. Current boundaries act as environmental boundaries to many species and the year-to-year variations in surface currents might affect the seasonal and life cycle migrations of pelagic and semipelagic species. The latter aspect has not yet been demonstrated with empirical data, although it is possible to do so with numerical modelling.

Advection of water by currents is an important factor which causes local changes in the environmental properties of the sea. Fish can be expected to respond directly to those environmental changes which are brought about by currents and, also, to respond to, or orient themselves directly to the current. Unfortunately, it is extremely difficult to observe directly the behaviour of fish in currents in the natural environment. Therefore, only relatively few *in situ* observations on the subject are available. For this reason it is necessary to draw upon knowledge from pertinent laboratory experiments. When so doing due attention must be paid to the fact that these experiments do not exactly reproduce the complex net of interactions of nature. Nevertheless, testing pre-established hypothetical models is a useful way of gaining more insight into the influence of currents on fish.

One would expect that currents affect the following aspects of fish distribution:

(*1*)   currents transport pelagic fish eggs and fry from spawning areas to nursery grounds and from nursery grounds to feeding grounds. Any anomaly of this ordinary transport can cause variations in the survival of any given year brood;

(*2*)   the migration of adult fish could be affected by a current serving as a means of orientation, and as modifier of migration routes;

(*3*)   the diurnal behaviour might be affected by currents (especially by tidal currents);

(*4*)   a current, especially at its boundaries, might affect the distribution of adult fish, either directly, through its effects on them, or, indirectly, through the aggregation of fish food, or by bringing about other environmental boundaries for them (*eg* temperature boundaries);

(5)   currents might affect the properties of the natural environment and thus determine indirectly the abundance of any given species and even the limits of its geographical distribution.

The following summary deals with the adult fish in relation to the currents. It attempts to summarize in a selective way the essential points of the past investigations on the subject and to establish some hypothetical models of the behaviour. These models need careful testing in nature before they can be accepted as valid.

The influence of currents on fish stocks is greatest in the egg and larval stages. Excellent investigations on this subject have been conducted by Walford (1938), Carruthers et al (1951) (who also suggest as a future approach the prediction of year-class strength from wind and current data), Rae (1957), Bishai (1960), and others. The effects of currents on the survival of larvae in combination with other environmental effects are described in Chapter 3 §3.1.

Current is perceived in the mechanoreceptor organ located at the lateral line of the fish. Lately it has been demonstrated that the 'heat exchange organ' in tuna is also located at the lateral line. It is possible that the 'heat exchange organ' also acts as mechanoreceptor.

Bull (1952), in his numerous carefully designed laboratory experiments on the behaviour of fish in relation to hydrographic factors, found no response of fish to the speed or direction of a current flow. Several other laboratory and field observations, however, have shown that fish respond and react to currents in several ways. Elson (1939) found that a sudden increase in the speed of current and in the degree of turbulence resulted in an increased activity of the speckled trout. The activity in still water took the form of random wandering, whereas the increased activity in a current was characterized by an active movement upstream. Brawn (1960) observed that herring respond to real and apparent currents greater than 3 to 9cm/sec by swimming upstream at a speed greater than the current speed until the maximum swimming speed is nearly reached.

Aleev (1958) analysed the manœuvrability of fish and their adaptability to water movements. He classified 25 species according to their horizontal and vertical manœuvrability and the coefficient of resistance for rectilineal movement. His observations indicate that different species react differently to the currents. Thus it is not possible to apply the results obtained with a given species to all fish. Probably there exists a lower threshold value of the current speed to which the fish respond and this threshold value varies from species to species.

Most field and laboratory observations show that fish usually head into the current. This is frequently the case even when they let themselves be carried along with the current. In weak currents other orientations can be observed.

Although several scientists have theorized that Pacific salmon use the sun for orientation in their migrations in the North Pacific, Favorite and Laevastu (1979) could reproduce numerically the known offshore distribution of Asiatic and American sockeye salmon when they assumed that the salmon swims against the surface current with a speed which is a function of the size of the fish and of the prevailing temperature of the water.

It has been observed that fish are inactive in very cold water and allow themselves to be carried along with the current. This is the case, for example, with herring during winter in the Norwegian Sea.

The orientation of fish (herring) within the shoal and the movements of the shoal as well might be regulated by currents. Some evidence for this can be drawn from the drift gill net fishery for herring. The fish go into the net in a given night usually from the same side and only on rare occasions from both sides.

It has been further observed that the light also plays a role in the orientation of freshwater fish to the current. Fish have been observed to swim against the current in light but to drift with it in darkness. The behaviour of adult fish in respect to the current (as well as in respect to other environmental conditions) is found to depend also on the physiological stage (eg on the maturity stage) of fish.

The behaviour of fish in relation to currents might be in most cases a result of a combined effect of several environmental factors. It is simple to correlate the diurnal behaviour of fish with the diurnal cycle of the light conditions which, when combined with the effect of currents (eg tidal currents) may considerably affect the distribution of fish. Davidson (1949) reported experimental results which indicate that salmon swim during the day and rest on the bottom at night, whereas eels burrow in the gravel in bright daylight, come out in the evening, and proceed rapidly at night. Thus, knowing the behaviour of a species in respect to light and knowing tidal currents (see further last part of this chapter), models of the drift of fish shoal can be conceived.

The activity of fish is also a function of the temperature of waters and thus it can be expected to affect both the migrations of fish and their transport by currents. It has been observed that in cold, near 0°C waters, fish are inactive and drift with the current. This fact is assumed in the migration of the Norwegian herring. During the early winter these herring aggregate in cold water pockets east of Iceland. Presumably they are carried into this convergence area by a branch of the East Greenland Current. When these cold water pockets are cut off into the warmer Norwegian Current waters and presumably sink, the fish move from the sinking cold water into the overlaying warm waters of the Norwegian Current and start their active migrations to the spawning grounds in the Norwegian coastal areas. The speed of migration is assumed to be a function of temperature (Devold, pers. comm.).

Trout (1957) described the migrations and movements of Bear Island cod. His description illustrates the interaction between currents, light, and seasonal migrations. He concluded as follows: 'Water movement is responsible for the changes in distribution of the Arcto-Norwegian cod. Yet, this is only a secondary effect of their changing behaviour in respect to changing light conditions which results in their primary annual vertical migrations and their annual depth range.

'Winter, rheotactic (responding to current), contranatant (swimming against the current) migration takes place on the bottom in the absence of light during the period of annual water transport maximum. Summer, passive, or denatant (drifting or migrating with the current) migration takes place during the lower summer peak of water transport at which time the cod shoals are largely pelagic in response to light. They are, thus, capable of being displaced horizontally, relative to the bottom, by movement of the water mass containing them. Maximum horizontal displacement is not achieved because a portion of the shoal spends a part of each 24 hours on the bottom. All known geographical limits of fishing are well within this maximum distance. Variations in summer transport values will affect the limits reached by shoals from year to year. Therefore, distribution is the resultant of interaction of changing behaviour with the cycle of movement of the water masses in which the cod are found.

'Normally, fish would tend to remain in this particular water mass, but transference from one water mass to another may take place as a result of changes in behaviour. In absence of water movement, horizontal movement will be of limited extent, but the annual depth range would be expected to persist.'

Fraser (1958) assumed that the diurnal migrations of fish (and plankton) also play their part in the transport of fish by currents. For example, organisms at the surface may drift in one direction at night, but in the daytime they may return with the deeper currents which might flow in the opposite direction.

A preliminary study of the boundaries of world ichthyofaunal regions in relation to currents indicates that sharp changes occur in the fauna where the speed of a permanent current along a coast exceeds one knot. This applies especially to the species inhabiting the down-current waters in the Agulhas, Kuroshio and Florida Currents, which have sharp distributional boundaries for less mobile species such as flatfish.

One of the first scientists to describe the possible relation between herring catches and tides was Tester (1938) who discovered an inverse relation between tidal difference and herring catches in British Columbia waters: 'At present, therefore, it would seem that the most plausible explanation of why herring are more available in the Swanson Channel fishery during the first and third quarters of the moon and less available during the new and full moon lies in the effect of tides on the movements of herring.'

Tester further assumed that strong tidal currents affect the transport of herring to and from various areas. Actually, the interaction between the diurnal behaviour and strong tidal currents might affect the transport of any fish. Let us assume that a fish species spends the daytime on the bottom and rises into the upper waters in the evening. Let us assume also that strong diurnal tides prevail in a given area. In these conditions the fish are expected to keep a given position on the bottom during day and are probably carried along with the strong tidal currents during the night. If the tides are diurnal, the ebbing current dominates during the night for about two weeks. The fish may be transported in one direction during this period and carried back again during the next period of two weeks.

Harden Jones (1957) studied the movement of herring shoals in relation to tidal

13

currents in the North Sea. His conclusions were: 'Echo surveys carried out in the Calais region in December, 1955, show that herring shoals move in the same direction as tidal currents. It is most likely that these observations were made on herring which had not yet spawned.

'It was not possible to come to any definite conclusion as to whether the fish were stemming the current, swimming with it, or being carried along passively. One set of results suggested that the herring were stemming the current at a swimming speed of 1 to 2 knots although they were being carried along the ground.'

Greer Walker *et al* (1978) studied the movement of plaice in the North Sea, using acoustic transponding tags. They found that: 'Plaice which moved more than 15km usually came off the bottom at slack water, moved downstream with the tide in midwater and returned to the bottom at the next slack water. When on the bottom the fish showed little or no movement during the opposing tide. The semi-diurnal (12h period) vertical movements were clearly related to the tidal cycle, ascents being more closely linked to slack water than descents. The regular pattern of behaviour – here called *selective tidal stream transport* – could provide a quick and economical means of movement for fish on migration through areas with strong tides.'

Several fishing methods rely on tidal currents. The most peculiar one is the so-called bag-net fishery off Bombay where the fish simply are carried by strong tidal currents into big anchored bag-nets. The fish seem to stem the current but, being relatively sluggish in the water with low oxygen content on the Bombay shelf, are transported with it.

An extensive summary of the effect of currents on the migrations of fish has been prepared by Harden Jones (1968), which should be consulted for any details of the effects of currents on fish.

2.3 Effects of light on fish

Summary
The light reception of fish is by eye and by pineal region which is located near the top of the brain. Most fish possess colour vision. Different species are adapted to different light intensities which can be regulated by depth selection. Some species are positively, some negatively phototactic and in some species phototaxis does not occur. Some of the photo-tactic reactions might be related to phototaxis in pelagic prey.

The 'hunters' need light for the location of prey, whereas the 'touchers' use vagile senses (senses of movement) for prey location. Feeding in most hunters occurs at relatively low light intensities during the morning and evening. Light stimuli affect the diurnal migration and shoaling behaviour in most species and thus affect the capture by gear.

Light is an important environmental factor in the lives of fish and other aquatic animals. Light affects fish directly through vision. Many fish rely on sight for capturing food, receiving signals that bring on or complete mating behaviour, locating shelter, and orientation. Light also affects fish indirectly through its influence on coloration. Light may also trigger and direct migrations and vertical movements, have a timing role in reproduction, and influence the rate and pattern of growth. Light perception in many fish occurs not only by way of the eyes but also the pineal region which is located near the top of the brain. The penetration of light and its absorption in the sea is described in Chapter 8. The influence of light on the physiology and behaviour of fish and its relationship to the availability of fish is illustrated in *Fig 4*.

The importance of light to the behaviour of fish and their food is clearly demonstrated by various fishing methods, known even at the most primitive stages of fisheries technology. Nevertheless, because of a certain parallelism in changes of light and temperature, their independent effects are not always clearly understood and separable. In this sub chapter special attention will be given to light as an environmental factor affecting fish.

Two good summary works on the effects of light on fish exist in the literature. Blaxter (1965) gave, from the applied point of view, an excellent summary of the effects of the changing light intensity on fish. He summarized the effects of light reception and especially the behaviour patterns in relation to light, such as feeding, shoaling, spawning, avoidance of gear, and optical attraction. Woodhead (1965) gave a summary of light upon the behaviour of demersal fish.

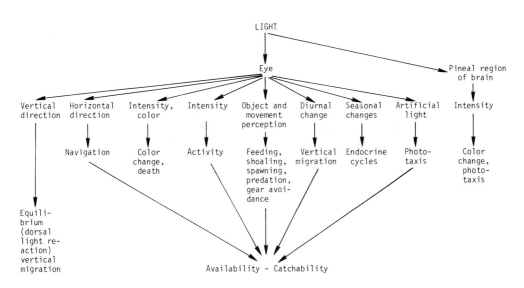

Fig 4   The influence of light on the physiology and behaviour of fish (adapted from Blaxter, 1965).

15

Fish are known to respond to light stimuli between 0·01 and 0·001 lux, depending upon previous adaptation to light or darkness. The lowest light intensities which bring about the maximum cone response in fish are found to be between 50 and 200 lux. In clear oceanic waters normal daylight can stimulate the human eye to a depth of about 250m, and it can be assumed that the fish which normally live at lower light intensities can use visual sense still deeper down. This is partly confirmed by the observations of Schärfe (1952), who found from echo-sounding observations that a certain lamp attracted fish at a distance of 26 to 28m, though for the human eye the lamp disappeared at a depth of *ca* 15m. This confirmation is valid if the turbidity of water is uniform from the surface to the depth of the lamp and below it. Clarke (1939) and Mohr (1960) found also that the 'visual capacity' of the fish eye corresponds roughly to the capacity of other higher vertebrates. However, owing to the turbidity of the water, the visual range frequently is a few metres only.

According to Bull (1952), Blinnius has found fish to possess a well-developed colour vision. Bull himself, however, failed to condition all fish visually in his numerous laboratory experiments on reactions of fish to environmental stimuli. The existence of colour vision in fish has been confirmed by the different reactions of fish to differently coloured nets. The fish can recognize the colour if the brightness of illumination exceeds a certain limit.

Some particular behaviour of fish in respect to light is frequently explained by the dorsal light reaction. Whereas animals orient at right angles to light, von Holst (1935) (from Blaxter and Parrish, 1958) found that *Crenilabrus rostratus* with both labyrinths removed would swim perpendicular to light from any direction, even if it involved swimming upside down in response to light from below. With the labyrinths intact, a fish would tilt its body in response to lateral light, but would not swim upside down. Schärfe (1952) found that fish dislike light coming from below and will disperse horizontally if descent is impossible.

Fish can be either positively or negatively phototactic, *ie* responsive to light. Many commercial fish are attracted to artificial light during the night, a fact which is utilized in practical fishing. Besides the direct phototaxis, the attraction to light may also be attributed to other causes. Zusser (1958) believes, according to his observation, that the night light acts as a signal for feeding. He observed that hungry fish seem more easily attracted by light than fish which do not feed (*eg* during the spawning season). Dragesund (1958) found that fish sometimes showed sudden upward movement towards the light when it was switched on (shock effect), but after a few minutes they either dispersed or packed and descended.

It appears from various observations and experiments that every species has a particular optimum light intensity where the activity of fish is at its maximum. Woodhead and Woodhead (1955) found that the activity of herring larvae was dependent on the light intensity. The lower threshold for positively phototactic herring fry adapted to light was 20 lux, and the maximum was at 4,000 lux. Activity decreased slowly in higher light intensities up to 65,000 lux. The darkness-adapted fry were still sensitive at 3 lux, and the optimum for the phototactic movement seemed to be around 100 lux. Imamura (1958) concluded that the equilibrum between light intensity and fish behaviour is dynamic where fish swim back and forth in certain limits of light intensity. He found that the optimum light intensity for carp is *ca* 0·2 to 20 lux.

The attraction of fish to artificial light is effective only with certain species. Amongst the positively phototactic fish can be mentioned young herring, sprat, saury, *etc*. Adult herring in the North Sea have shown little aggregation in response to artificial light. Balls (1951) found that adult herring avoid high light intensities and have diurnal negative phototactic migrations, *ie* they go deeper in daylight. Cod, hake (*Merluccius merluccius*) and ling (*Molva molva*) are also slightly phototactic, and it is generally known that both Atlantic and Pacific eels are negatively phototactic. Kawamoto (1958 and 1959) found in addition that parrot fish, file fish, seer fish, tiger puffer, and barracuda are positively phototactic and are especially attracted to blue and green lights. However, the colour effect may be explained by the fact that blue and green lights penetrate deepest into the water.

The effect of artificial light on fish is also found to be influenced by other environmental factors and varies in some species with the time of day. It is also known that after hatching, whitefish larvae (*Coregonus spp*.) leave the rivers during the night. Experiments in glass tubes with running water show that the small larvae of whitefish can swim against relatively strong currents in the light. However, as soon as the light is switched off, these larvae drift with the current.

Kojima (1957) found in laboratory experiments that young pilot fish (*Naucrates ductor*) and parretas (*Oplegnathus fasciatus*) react to floating substances of various colour. This reaction might also be connected with tactile senses, because certain fish (*eg* herring in captivity) have occasionally been observed to swim along barriers in darkness. It has been demonstrated by several researchers that pelagic fish are attracted to floating objects. Hunter and Mitchell (1968) found that small pelagic fish are attracted first and are followed by larger fish. The residence time of skipjack and yellowfin tuna near floating objects was about ten days.

Hasler *et al* (1958) concluded from field and laboratory experiments that the sun might serve as a reference point for fish in their migrations and in the search for their spawning grounds.

Feeding of those fish species which find their prey by sight, the 'hunters', is often affected by light conditions. The filter feeders and the 'touchers' can, of course, locate their prey in complete darkness. Also, the tactile sense might be used by fish to find vagile (moving) prey. Hempel (1956) found that plaice were feeding during the day on bottom animals. Those few fish which feed during the night were feeding on pelagic polychaetes (marine worms). He also found that although lemon sole (*Microstomus microcephalus*) feed mainly during the night on vagile animals, like shrimps and other crustaceans, in some areas they also eat during the day.

2.4 Influence of other environmental factors

Summary

Oxygen content in the marine environment does not affect the fish in normal conditions as the dissolved oxygen content in the oceans varies within relatively narrow limits. Exceptions are those specific conditions such as oxygen-minimum layers below tropical thermoclines and deeper holes in the Baltic Sea where oxygen content of the bottom water is low.

Olfaction sense is assumed to be well developed in fish and is used by salmon for homing. However, little is known on reaction of fish to olfactory stimulus.

Salinity of the water affects the osmoregulation in fish and has a large influence on fertilization and development of the eggs. Different species are adapted to different salinities. Some species are euryhaline (adaptable to wider range of salinities), but most are stenohaline (with relatively narrow salinity tolerance limits).

Waves and turbulence affect at least some fish and influence their vertical movements and distribution.

Many fish produce sound and react to some sound. The effects of sound on fish behaviour is, however, little investigated.

Aquatic respiration requires the uptake of oxygen from the water and the elimination of waste carbon dioxide. The gills are typically the site where this gaseous exchange takes place although in some fish gas exchange with the water takes place through the skin.

Ordinarily the rate of oxygen consumption by fish may be taken as a general measure of the intensity of its metabolism. This rate is influenced by the size of the fish and the characteristics of the water such as temperature and carbon dioxide content.

The activity of which a fish is capable is correlated with the difference between its resting (standard) and active rate of metabolism, and is dependent upon the oxygen content of the water. Thus a reduction in the oxygen content of the water reduces the active rate of metabolism and therefore restricts the activity of the species. Activity in this respect includes development, growth, and movement. *Figure 5* illustrates schematically the influence of the oxygen content of water on the activity of fish and its relation to their future abundance.

The oxygen content of the water varies with temperature and with depth. In major parts of the surface layers of the oceans the oxygen content varies within relatively narrow limits. However, below thermocline layers, near the bottom and in some tropical regions oxygen content can be very low and severely affect fish and other marine life. An example of the effect of fluctuating and low oxygen content of the water on the growth of brook trout is shown in *Fig 6*. *Figure 7* shows the influences of oxygen concentrations and temperature on the swimming speed of juvenile coho salmon.

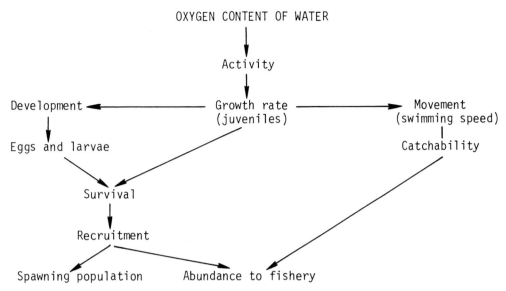

OXYGEN CONTENT OF WATER

Activity

Development ← Growth rate (juveniles) → Movement (swimming speed)

Eggs and larvae

Catchability

Survival

Recruitment

Spawning population   Abundance to fishery

Fig 5 Schematic illustration of the influence of oxygen content of water on the activity of fish and its relation to population abundance.

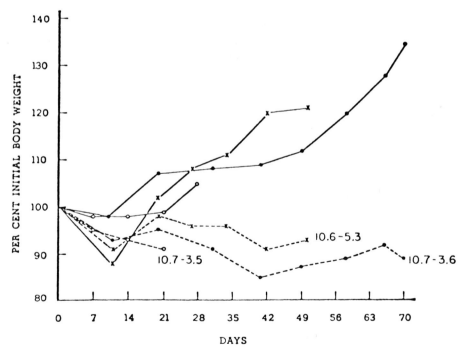

10.6 - 5.3

10.7 - 3.5

10.7 - 3.6

DAYS

Fig 6 Growth of yearling eastern brook trout, *Salvelinus fontinalis*, at constant high (solid lines) and various daily fluctuating levels of oxygen (broken lines). Numbers indicate upper and lower levels of oxygen in mg/l (from Whitworth, 1968).

19

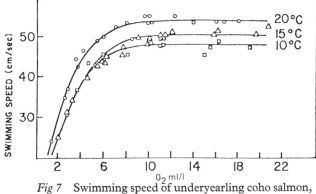

Fig 7 Swimming speed of underyearling coho salmon,
Onchorhynchus kisutch, in relation to oxygen concentration
(from David *et al*, 1963).

Under normal conditions in the sea, the dissolved oxygen content of water does not become a limiting factor in the distribution of fish. However, in some cases the oxygen might affect the behaviour of fish. Johansen and Krogh (1914) found that oxygen deficiency in the water plays a role as a retarding factor in the development of plaice eggs.

Chemical reception in fishes is divided into three categories: olfaction or smell, gustation or taste, and a general chemical sense. Odour perception is generally regarded as a distance perceptor, and its keenness is considerably greater than taste perception. In fish, smell and taste are both brought about by aqueous solutions so that the distinction is made anatomically and physiologically. The general chemical sense in fishes refers primarily to the free nerve endings in the skin of fishes which are presumed to have chemical sensory capabilities. Reactions following stimulation of the general chemical sense are usually negative or defensive.

Chemical reception in fishes is involved in the procurement of food, recognition of sex, defence against predators, avoidance of dissolved substances, parental behaviour, and in orientation (*Fig 8*).

The volume of water and concentration of salts in the internal body fluids of fish and shellfish are influenced by the salt concentration of their environment. Aquatic marine animals faced with a diversity of environments have selectively evolved a regulatory system to control these factors within very narrow limits. This control is known as osmoregulation and is one of the primary functions of the gills and kidneys in fish. Osmoregulation requires the expenditure of energy by the organism, the amount of which depends upon the difference in salt concentration existing between the external environment and internal body fluids.

Salinity tolerances and preferences in marine organisms vary with their life history stage, *ie* eggs, larvae, juveniles, and adults. Salinity appears as an important factor influencing the success of reproduction in some fishes and the distribution of various life history stages (*Fig 9*). In addition, salinity gradients may function in orientation during the migration of such species as Pacific salmon.

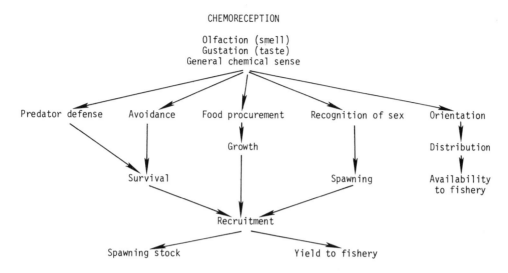

Fig 8 Schematic illustration of the function of chemoreception in the behaviour of fish and its importance in influencing the abundance and availability of fish.

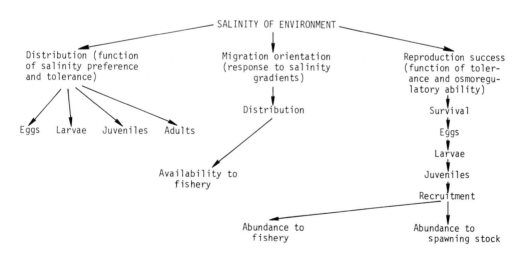

Fig 9 Influence of environmental salinity on the distribution and abundance of fish and shellfish.

Salinity variations in offshore areas are relatively small, but in coastal areas are considerably larger due to variations in river run-off. These variations affect the osmotic regulations of fish and determine the buoyancy of pelagic eggs. Although there are numerous articles describing the apparent relations between fish behaviour, or distribution, and the salinity of the water, those relations are not necessarily direct. The variations of salinity very often indicate the change in water masses or in their stability conditions, including vertical mixing. Thus the observed correlations might be brought about indirectly by the advective effects.

The direct influence of salinity on most fish species can be considered minor only as the salinity in the open ocean varies only between about 30 and 36‰. This has been demonstrated by Holliday and Blaxter (1960) (*Fig 10*). They found that fertilization, development, and hatching of herring eggs occurred in salinities ranging from 5·9 to 52·5‰. The salinity tolerance of both spring- and autumn-spawned larvae was found to lie between 2·5 and 52·5‰ for 168 hours. They concluded that it seems unlikely that the osmotic forces will constitute a serious hazard to the herring larvae. (In this case with the 'osmotic force' is meant the tendency between sea water and the liquids in living cells to penetrate the cell membranes. This tendency varies with the salinity of sea water.) Further laboratory investigations are needed to find out the general validity of this conclusion based upon studies with herring larvae. Furthermore, the limits of the optimum salinity range, especially without any adaptation period to exceptional conditions, must be for any fish species much narrower than the utmost range of tolerance.

One indirect relation between salinity and fish behaviour is described by Seckel and Waldron (1960) who investigated the occurrence of skipjack (*Katsuwonus pelamis*) in Hawaiian waters. It was recognized that the good fishing years for skipjack occurred when the islands were bathed in California Current Extension waters characterized by

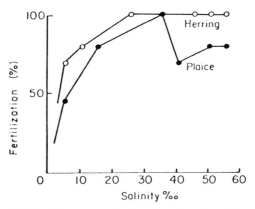

Fig 10 Percentage fertilization of herring, *Clupea harengus harengus*, and plaice, *Pleuronectes platessa*, eggs (from Holliday, 1969; data for herring from Holliday and Blaxter, 1960).

22

salinities lower than $34\cdot7\%_{oo}$. Lower catches were associated with the Western North Pacific waters with salinities higher than $35\cdot0\%_{oo}$. It was further observed by them that the time of the northward movement of the boundary between the California Current Extension and Western North Pacific waters is related to the annual skipjack landings. Whenever this movement occurred in February, which implies a well-developed California Current system, average or better than average fishing year resulted. These relations allow the prediction of fishery to be made several months ahead.

The fact that fish avoid the upper layers of the sea during heavy weather is well known to fishermen. Also, onshore-offshore movements of fish are, to a certain extent, affected by waves. As investigations of Kamiura (1958) and Nakai (1959) indicate, some species might be more sensitive to waves and marine noises than others. Obtaining quantitative data on the influence of waves on the vertical movement of fish is extremely difficult, because these movements are influenced also by other factors, especially by light conditions.

Harden Jones and Scholes (1980) noted that plaice in the North Sea moves into deeper water during heavy seas. They suggest that this movement might be caused by fish seeking quieter water where the turbulence caused by waves is reduced. They consider it unlikely that plaice get seasick, but report instances where fish held in the tanks on board vessels were seasick.

The detection of sound by fishes is briefly mentioned here because it is one of the several senses in fish which may be utilized in the identification of and orientation to currents and therefore might have an influence on fish distribution and availability. Three physiological mechanisms are employed by fish to discriminate sound: the inner ear, the swim bladder, and the lateral line system. The function of each of these mechanisms in hearing is covered in detail in several texts on the physiology of fish. The effects of sound on the behaviour of fish have been but little investigated in the past.

3

Effects of environment on the life history of fish and on the behaviour of the stocks

This chapter gives a review of the influence of environment (and variations in it in space and time) on the reproduction, survival, and distribution of fish. Only a few examples of the interactions between the environment and fish stocks are given, few of which can be generalized to other species and areas, as the environmental adaptions and requirements vary from species to species.

The whole life history of fish species is closely dependent upon the environmental conditions. It can be postulated, therefore, that application of oceanographic knowledge to fisheries problems can result in better understanding of relations between the behaviour of fish stocks and their environment and will promote the formulation of fisheries prediction systems.

It should be pointed out that most of the earlier fish-environment studies have analyzed only the effects of one environmental feature in relation to fish behaviour and distribution at a given time. The interaction of the fish and the environment, however, is extremely complex and it is obvious that a number of features and processes will jointly affect fish populations. One of the recent, more thorough analyses of the collapse of Icelandic and Atlanto-Scandian herring stocks in the late 1960s by Jakobsson (1978) shows the complexity of various environment-stock interactions (which are described in greater detail in Chapter 4). He concludes that the adverse environmental conditions in the late 60s must have affected the herring in various ways and almost certainly contributed greatly to the collapse of the stock, directly or indirectly.

A generalized schematic presentation of the effects of environment on the stocks and processes affecting the stocks is given in *Fig 11*.

3.1 Reproduction and recruitment in relation to the environment
Summary

The effect of environment on fish and fish stocks is most pronounced during spawning and during egg and larval development when recruitment can be affected by environmental anomalies. Different species and even different stocks have relatively narrow temperature preference limits during spawning. Temperature anomalies during

24

Stock condition and process	Environmental factors affecting the stock
REPRODUCTION, LARVAL SURVIVAL	
Reproduction potential	
Number of spawners, maturation	Temperature (effect on maturation, displacement of spawning) Food availability (starvation, effect on maturation)
Survival of eggs, larvae	Grazing on eggs, larvae (abundance of predators) Availability of proper food Dispersal (transport, mixing)
PREFISHERY JUVENILES	
Abundance,distribution	
Transport, migrations	Currents
Grazing, survival	Availability of food Presence of predators
Growth	Temperature (especially during winter) Availability of food
EXPLOITABLE STOCK	
Recruitment from juveniles	Same processes as affecting abundance and distribution of juveniles
Distribution, abundance	
Grazing by mammals	
Availability to fishery	Temperature anomalies (affecting seasonal migration) Availability of food (preferred prey)

Fig 11 Schematic presentation of the environmental factors affecting the stocks and processes in them.

spawning seasons can delay and/or hasten spawning and cause displacements of spawning grounds. The time required for egg development and hatching is a function of temperature in all species.

The survival of larvae is dependent on the availability of proper planktonic food at the right time and place. This availability is affected by environmental anomalies and by anomalous transport of eggs and larvae. The occurrence of storms and associated turbulence during hatching and early larval periods are known to cause heavy mortalities of embryos and early larvae.

The year-class strengths in most fishes are determined by factors other than the size of spawning stock. These other factors include predation and various environmental influences. In some species (*eg* sole) exceptionally cold winters produce good year-classes, which is thought to be brought about by decreased predation. Comparison of year-class sizes of many species indicates that no one single environmental factor can be responsible in all cases for year-class strength variations.

The influence of water temperature on fish behaviour is most pronounced during spawning. But also temperatures prior to spawning are highly significant since they influence the ripening of the sexual products. Hodder (1965) described the influence of temperature, at certain critical periods during the initial development and early maturation of the ova, on the fecundity of haddock. Poulsen (1944) stated that cold months preceding the spawning of cod (*Gadus callarias*) may have a considerable effect in delaying the maturation of gonads. Every stock of fish has a 'normal' temperature range, possibly with a seasonal cycle. At temperatures below this range the ripening of the sexual products is delayed, and the opposite is the case at higher temperatures. Therefore, the past history of the temperature is decisive to the arrival of a stock to its spawning grounds. The theory of Devold (1959), which provides an explanation of the fluctuation in the stock and of the year-to-year shifting of spawning grounds of the Atlanto-Scandian herring (*Clupea harengus harengus*), is partly based upon similar conclusions.

Jean (1956) summarized the water temperatures during the spawning of herring and showed that herring spawn over a wide range of temperatures in different localities. *Table 1* shows that water temperatures at which herring spawn range from 0 to 12°C in the spring and from 8 to 15°C in the autumn. Actually, the mean spawning temperature for spring is $6 \cdot 5 \pm 2 \cdot 1$°C. In *Table 2* a few water temperatures are given at which various other fish spawn and thrive in different localities. The seasonal south-north progression of spawning time and the temperature for spawning of the herrings, *Clupea harengus* and *pallasii*, in NW Atlantic and in NE Pacific, respectively, are shown in *Table 3*. These data indicate that the mean spawning temperature for spring is $7 \cdot 4$°C and for autumn $10 \cdot 7$°C, both of them well fitting into the above pattern.

Water temperature may hasten or delay the onset of spawning of Atlantic herring as shown by Lauzier (1952). He observed that mean water temperatures at Magdalen Islands, Gulf of St. Lawrence, from 1933 to 1950 were $0 \cdot 3$°C in April and $4 \cdot 5$°C in May. During these years, the catches of spawning herrings were less abundant in April than in May in a proportion of 1 to 18. In 1951, mean water temperatures at Magdalen Islands were $3 \cdot 8$°C in April and 6°C in May. In contrast with previous years, the catches of spawning herring were more abundant in April than in May in a proportion of $4 \cdot 5$ to 1. Spring spawning appears to have taken place earlier in 1951 because of temperatures higher than normal in April.

The correlation between water temperature and spawning of marine animals is generally recognized. Orton (1919) and Hutchins (1947) have shown that several species of marine invertebrates spawn within narrow limits of temperature. Runnström (1927) correlated the geographical distribution of plaice (*Pleuronectes platessa*) with its temperature requirements at spawning. Allen (1897) and Sette (1950) have shown that mackerel (*Scomber scombrus*) spawns at temperatures ranging from 12° to 15°C. Spawning of this species takes place earlier in the southern than in the northern part of its range.

The effect of water temperature on spawning has been studied in detail by various workers. Mankowski (1950) stated that low temperatures during the spawning season delayed the spawning of Baltic cod, whilst higher temperatures hastened spawning. Dragesund (*cf* Devold, 1959) has kept living Atlanto-Scandian herring in nets in

Table 1
Water temperatures during the spawning of herring (*Clupea harengus*) in different areas
[Rearranged from a table by Jean (1956)]

Locality	Season (Spring)	Temperature (°C)	Season (Autumn)	Temperature (°C)	Authors
Norway	II–III	3·0–7·0			Buch (1885) Johansen (1924) Runnström (1941)
North Sea			VIII–XI	6·0–13·0 12·0–14·0	LeGall (1935) Johansen (1924)
Kattegat	I–VI	4·0–12·0			Johansen (1924) LeGall (1935)
			IX–X	11·0–13·0	Johansen (1924) LeGall (1935)
Baltic Sea	IV–VI	6·0–11·0			Altnöder (1929) LeGall (1935) Cieglewitz and Posadzki (1947)
			IX–X	11·0–14·0	Hessle (1925) Cieglewitz and Posadzki (1947)
English Channel	XII–II	6·0–12·0			Fage (1920) Johansen (1924)
Scotland	II–III	3·9–7·1			Fulton (1906) Wood (1936)
			VII–IX	11·0–12·0	Fulton (1906) Fage (1920) Johansen (1924) Wood (1936)
Barents Sea	II–IV	0·0–6·9			Rass (1926) Rass (1939) Manteufel and Marty (1939)
Iceland	Spring	5·0–9·0			Taning (1936)
Faeroes	III–IV	*ca.* 3·0			Johansen (1921)
Grand Manan			Autumn	8·0–11·0	Bigelow and Schroeder (1953)
Cape Cod			Autumn	11·7–12·8	Bigelow and Schroeder (1953)
Newfoundland	Spring	8·0			Tibbo (1946)
Block Island Sound			Late summer	13·0–15·0	Merriman and Warfel (1948) Merriman and Sclar (1952)
Magdalen Islands	IV–V	3·8–4·5			Lauzier (1952)
Mean	Spring	6·5 ±2·8	Autumn	11·2 ±2·1	

Table 2
Optimum temperatures for fish in different areas

Species	Locality	Optimum temperature range (°C)	Spawning temperature (°C)	Remarks	Author
Cod (*G. callarias*)	Bear Island, Spitsbergen	2–4			Lee (1952, 1956)
,, ,,	Newfoundland	(3—) 5–7			Le Danois (1934)
,, ,,	Newfoundland	0·5–7			McKenzie (1934)
,, ,,	West Greenland	(2·5—) 3–4		Best catches with surface and bottom long line	Rasmussen (1955)
,, ,,	Motovsky Bay		0·4–2		
,, ,,	Northern Norwegian waters		2·5–5		Pertseva (1939)
,, ,,	Skagerrak		4–6		
,, ,,	Newfoundland		3–5		
,, ,,	Newfoundland, Spring	0–3			
,, ,,	Newfoundland, Summer	3·5–5·5			Thompson (1943)
,, ,,	Icelandic waters		> 6		Schmidt (1926)
,, ,,	Lofoten		3–6·5		Sund (1935)
,, ,,	Danish waters		3–7		Poulsen (1931)
,, ,,	Nova Scotia Banks, Spring	—0·5–1·5			McKenzie (1934)
,, ,,	Nova Scotia Banks, Summer	2·5–5·5			
,, ,,	Labrador waters	> 2		Poor catches if temp <1·5 or > 4·0	Rasmussen (1955)
,, ,,	Eastern Atlantic, Winter	2–3			Lee (1952)
,, ,,	Eastern Atlantic, Summer	3–5			
,, ,,	Western Atlantic	3–5			Le Danois (1932)
,, ,,	Hamilton Inlet Bank, Spawning (April–May)		2·5–3·1		Templeman and May (1965)
	Summer	—0·2–8·7			
Haddock (*M. aeglefinus*)	Western Atlantic	5–7			Le Danois (1932)
,, ,,	Off Eastern Canada, Winter	3–6		Greatest abundance	McCracken (1965)
	Summer	6–8		,, ,,	
,, ,,	Grand Banks, Feb–June	2–8			Templeman and Hodder (1965)
Plaice (*P. platessa*)	North Sea		4–7		Kändler (1955)
Shad (*A. sapidissima*)	York River	7–15			Massman and Pacheco (1957)
Baltic herring (*C. harengus harengus*)	Gulf of Riga	8–12			Bérziņś (1949)
Baltic sprat (*C. sprattus*)	Gulf of Riga	10–15			

Table 2
Optimum temperatures for fish in different areas

Mackerel (*S. scombrus*)	North Atlantic		12–15		Allen (1897) Sette (1950)
„ „	North Atlantic		10–15	Salinity 26–33	Jensen (1955)
„ „	North Atlantic	12–14		Avoids temp <4	Dannevig (1955)
"Iwashi" sardine (*S. melanosticta*)	Sea of Japan	12–16			Uda and Okamoto (1936)
„ „	Sea of Japan		13–17 Optimum: 14–15·5		Uda (1959, d)
Pacific sardine (*S. caerulea*)	Off California		13–17 Optimum: 15–16		Ahlström (1959)
Pilchard (*S. pilchardus*)	English Channel		9–16·5		Cushing (1957)
S. African pilchard (*S. ocellata*)	Off South-West Africa		14·9–19·6		Mathews (1959)
Tuna (*T. thynnus*)	Kuroshio region	14–18 18–20		Favourable fishing Best fishing	Kawana (1934) Uda (1952)
Anchovy (*Engraulis*)	Mediterranean Atlantic	6–29	13–29 10–23 Optimum: 13–17·5		Various sources Baxter (1967)
„	Off Japan		11–29		Hayasi (1967)
„	Off Argentina		10–17 Optimum: 10–17		De Ciechomski (1967)

Table 3
Time and water temperature for spawning of Pacific and Atlantic herrings of North America
(from Scattergood, Sindermann and Skud, 1959)

Source	Location	Spawning time	Temperature (°C)
Pacific (*Clupea harengus pallasii*)			
Miller (1956)	California	Jan–April	8·0–10·0
Westerheim (communication)	Oregon	Jan–April	3·8–12·3
Outram (1955)	British Columbia	Feb–April	4·4–10·7
Skud (ms.)	Southeastern Alaska	April–May	6·1–11·0
Rounsefell (1930)	Western Alaska	April–May	3·0–5·5
Atlantic (*Clupea harengus harengus*)			
Tibbo (1956)	Newfoundland	May–June	8·0
Jean (1956)	Quebec–Bay of Chaleur	May–June	2·2–12·4
Bigelow and Schroeder (1953)	New Brunswick–Grand Manan	Aug–Sept	7·1–11·1
Bigelow and Schroeder (1953)	Massachusetts–Cape Cod	Oct–Nov	up to 12·8
Tibbo, *et al*	Georges Bank	Sept–Nov	11·9–13·9

shallow water. Thus, throughout the spring these fish were forced to stay in the cold surface water with the result that they did not spawn until June when the temperature of their abnormal environment started to rise. This experiment shows that spawning can be delayed for three months by keeping the fish in water colder than the optimum. The narrowness of the temperature range suitable for spawning [in a few cases different for different distribution of the same species (Pertseva, 1939)] affects the geographical distribution of spawning. For example, the cod spawn in Motovsky Bay three weeks later than near the Lofoten Islands and one and a half to two months later than in the North Sea. Fish appear to respond to their oceanographic 'climate' rather than to geographically fixed reference points during the spawning season (Ahlström, 1959). Therefore, unusual temperatures on the spawning ground during the spawning season force the fish to spawn in other areas than those to which they normally go for this purpose (Simpson, 1953). Long-term temperature changes may therefore cause the periodic northward and/or southward displacements of spawning (and fishing) grounds. The conclusions of Nikolajev (1958) on the behaviour of the Baltic herring are in accordance with these statements: 'The abundance of the spring and autumn Baltic herrings seems to fluctuate within different ranges, the abundance of the spring herring increasing during cold-water periods (according to the winter regime 1926-30) and decreasing during warm-water periods (1931-39 and 1951-57), the abundance of the autumn herring, on the contrary, increasing during warm-water and decreasing during cold-water periods. This fact seems to be consistent with the geographical distribution of these particular herrings and of all spring and autumn herring races in general. The spring herrings prevail in the sub-arctic waters of the Norwegian, Barents, and White Seas, as well as in the Northern Baltic, whereas the autumn herrings mostly occur in the boreal regions (northerly waters) of the North Sea and in the Southern Baltic. The main cause of these differences seems to lie in the fact that biological peculiarities (particularly the spawning season) make for a better utilization of the feeding resources of colder waters by the spring herring and of warmer waters by the autumn herring.' (cf Devold, 1959.)

The development of the eggs and larvae is undoubtedly the most critical period in the life history of fish, because this is when they are most strongly influenced, directly and indirectly, by physical conditions.

Temperature directly influences the rate of development and, in conjunction with salinity, will determine the prevailing water density, thus affecting the buoyancy of the eggs. The length of time taken for the incubation of eggs, as well as the length of larval life, depends directly upon the temperature of the environment.

Jean (1956) has surveyed the literature on the incubation period of Atlantic herring eggs at various temperatures, based partly upon laboratory experiments, partly upon observations in the sea. These data, from 21 different sources, are plotted in *Fig 12*. The probable form of the relationship between water temperature and the incubation period is exponential, with an asymptote above zero. The best fitting curve is expressed by the equation:

$$T = 4 + 44 \cdot 7 \exp(-0 \cdot 167 \vartheta) \tag{1}$$

For the computation of this empirical equation the data of Hesse *et al* (1937, from Gunter, 1957) were also used, according to which herring eggs develop in forty to fifty days at 0·5°C and in six to eight days at 16°C. The above equation indicates that the incubation period is about forty nine days at 0°C, and is always four days or more regardless of the temperature.

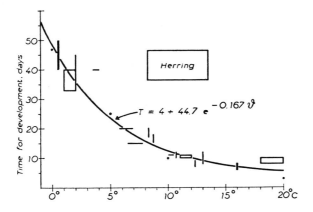

Fig 12   Incubation period of herring eggs at various temperatures. Observations by Hesse (from Gunter, 1957) are indicated by heavy lines; other observations, collected from different sources by Jean (1956), are indicated by thinner lines.

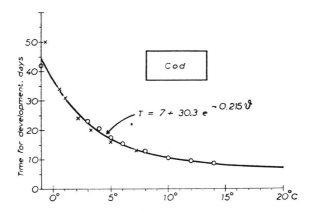

Fig 13   Incubation period of cod eggs at various temperatures (x Earll 1880; o Dannevig 1895).

31

Correspondingly, the incubation period, as a function of environmental temperature for cod eggs, based on data given by Earll (1880) and Dannevig (1895), is as given in *Fig 13*. In this case the best fitting curve is represented by the expression:

$$T = 7 + 30 \cdot 3 \exp(-0 \cdot 215\vartheta) \qquad (2)$$

which means, among other things, that the incubation period for eggs is thirty seven days at 0°C, and is always seven days or more regardless of the temperature. It is interesting to note that, according to this treatment of the data, the incubation time for both herring and cod eggs is about eight and a half days if the water temperature is 13·5°C.

Ito (1958) gives data on the incubation time for sardine (*Sardinops melanosticta*) eggs. The best fitting curve is expressed by the equation:

$$T = 0 \cdot 5 + 28 \cdot 8 \exp(-0 \cdot 159\vartheta) \qquad (3)$$

which means that the hypothetical minimum incubation period for sardine eggs is approximately half a day. At 10°C the period is six and a half days (*Fig 14*).

The prevailing water temperatures – and currents – during and after spawning are considered the most important factors determining the 'brood strength' and survival of larvae of the commercially most important fish species (Rounsefell, 1930; Uda and Honda, 1934; Kurita, 1959). This postulate has been used in the prediction of the strength of coming year-classes (Chase, 1955). In medium and higher latitudes, high temperatures coincide with rich year-classes in most species, and a low temperature with poor year-classes (Täning, 1951; Hermann, 1951).

Actually there are several other ways in which the temperature can influence the survival of larvae. The most important of these is probably its effect on the availability of food. It is evident that the availability of food suitable for larvae at the appropriate time

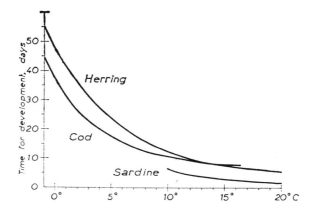

Fig 14   Incubation periods of herring, cod and sardine eggs.

32

is related to the phytoplankton production, which in turn is closely related to the seasonal changes of temperature and to the amount of sunlight (which normally shows a close correlation with the sea water temperature). Also the abundance of zooplankton, important as food for the larvae, is related (*a*) to the abundance of phytoplankton, and (*b*) to the spawning period of adult planktonic animals, which again is controlled by temperature. Many authors have reported that warmer water during and after spawning has resulted in better year-classes of cod because of the consequent greater availability of plankton as food (Wise, 1958). In general, too high or too low temperatures may put the development of larvae 'out of phase', so that larval development occurs before or after the peak population of the proper plankton.

It is worthy of mention that the influence of sea water temperature on the year-class strength may also be exerted through competition, as suggested by Marr (1959): 'The lack of spawning success off southern California is attributed to environmental conditions, especially to the below average temperature regime which has delayed sardine spawning off southern California by one or two months. It is proposed that under these conditions sardine larvae are unable to compete successfully with the anchovy, which has a lower temperature threshold.'

There are relatively few tests of the predictive value of temperature in respect to area and times of spawning. Myberget (1965) found that the spawning of mackerel started in 1957 and 1958 in the last days of May, whereas spawning started before mid-May in 1959. Sea surface temperatures in 1959 were warmer than in the two previous years. The spawning of mackerel takes place in water temperatures of 10 to 16°C. Neller (1965) demonstrated the influence of water temperature and wind on the spawning, migrations, and catches of a local herring population and showed that the environmental data can be used for predictive purposes.

Laboratory studies on the embryonic development of Argentina anchovy (*Engraulis anchoita*) by De Ciechomski (1967) gave as hatching time 68 to 72 hours at 14-15°C and 50 to 53 hours at 19-20°C. Spawning can normally take place in temperatures between 10° and 17°C, 4°C is lethal. Salinity tolerance is from 26 to 50‰ and light has an influence on hatching. De Ciechomski found that the embryos are very sensitive to mechanical factors. He concluded therefore that storms and heavy seas can destroy developing embryos.

Although it was shown in the late 1930s by Rollefsen that storms and high turbulence in the water can be lethal to cod eggs and embryos, no follow-up studies, besides that by De Ciechomski, have been undertaken on this subject.

It has been generally accepted that there is a considerable variation of spawning success that cannot be associated with the size of the spawning stock (Murphy, 1966) and that catches vary and depend upon the size of the year-classes. Many of these variations are due to the influence of environment during hatching and early stages of larval development.

Light may have various real – and apparent – effects on the spawning of fish and on the larvae. Sullivan and Fisher (1953) postulated that the amount of light available can influence the maturity time of the fish. This is possibly a physiological effect and regulates the spawning time in several species of fish so that spawning takes place during the most favourable temperature conditions and hatching during the period of most abundant food. Nevertheless, the indirect effects of temperature are usually difficult to separate from the light effects. In spite of this difficulty, it is worthwhile mentioning that Allen (1909) found that the amount of light in early spring affects the survival of fish larvae. It is probable that in this case the effect is purely indirect as the amount of organic production, and thus availability of food, depends upon the availability of light energy.

Light also affects the behaviour of larvae. Bridger (1956) compared the day and night catches of herring larvae with the Heligoland larvae nets. Over a series of observations this net took six times as many herring larvae and four times as many pilchard larvae by night as it did by day. The increased ratio of the night catches was not the same for all stages of development. He concluded that the limited day catches were brought about by the effect of daylight. This conclusion could be verified by means of high-speed plankton samplers. Woodhead and Woodhead (1955) found that herring larvae exhibited a positive phototactic movement, their activity being proportional to the light intensities. The lower threshold of the light-adapted larvae was about 20 lux and the maximum activity appeared at 4 000 lux. At lower light intensities, down to 3 lux, and also in red light, the darkness adapted larvae swam only vertically upwards when active, or sank passively. At higher light intensities (about 100 lux) these larvae swam also horizontally during their activity period.

The vertical migration of pelagic larvae could be attributed also to the fact that several phytoplankton organisms produce, during photosynthesis, substances poisonous to animals. Therefore the downward movement of pelagic larvae during the morning hours may constitute a mechanism of escape from the influence of these substances, while the upward movement of the evening hours could be considered simply a feeding migration.

The effects of environment on the spawning success and recruitment vary, not only from species to species, but also from one stock of a species to another stock of the same species. This has been demonstrated by Nakai *et al* (1968) through the investigation of the effects of the extraordinarily cold water masses that covered the sea areas surrounding Japan in early 1963. The low temperature caused widespread delay of the spawning activity and southward shift in distribution of the sardine, anchovy, mackerel, and jack mackerel along the Pacific coast of Honshu. However, changes in distribution and abundance differed depending upon species and stock. The sardine suffered the most disastrous damage because the cold water hit the major spawning ground confined in the waters around the Boso Peninsula in 1963. The anomaly caused only limited effects on the anchovy through delay of the spawning season because of the large stock size and wide extension in space and time of the anchovy spawning activity. The mackerel stock in the Pacific waters along Honshu comprises two stocks; Honshu-Pacific stock and Kyushu-Pacific stock, off southern Kyushu. The former stock has dominated over the latter on the fishing ground. The low temperature decreased reproduction of the Honshu-Pacific stock while it increased recruitment of the Kyushu-Pacific stock through shifting the spawning ground southward. The jack mackerel of the 1963 year was abundantly landed on the Pacific coast because of a southward shift of spawning during the cold anomaly.

Holden (1978) found that in the North Sea large year-classes of sole invariably follow winters in which sea temperatures fall markedly below average. The year-class of plaice following the severe winter of 1963 was three times the average, but except for this there is no association between year-classes of plaice and severe winters. Comparison of year-class sizes of other species showed no common feature that points to one single factor being responsible for the high level of recruitment to the cod, plaice, sole, haddock, and whiting populations in the sixties in the North Sea.

Hempel (1978a) concluded that for haddock and whiting year-classes in the North Sea, a positive correlation in the abundance of the two can be found, pointing to the likely effect of a combination of favourable environmental influences on survival and growth of early life history stages. It is possible that this correlation can also be brought about by predation. Dragesund and Nakken (1973) found that in most years the year-class strength of the Norwegian spring-spawning herring was determined by factors other than the size of spawning stock, and that the predation on larvae by haddock and saithe accounted for a great part of the reduction of the number of larvae.

Studies on the effect of environment on spawning and early recruitment of fish show that a great multitude of environmental factors can be involved and that the effects are greatly variable in space and time and from one species to another.

3.2 Influence of environment on the survival and mortalities

Summary

Mortalities may be caused by abnormal and rapid temperature changes, even when temperatures are considerably above freezing and the acclimatization temperature of the species is relatively high. The dynamics of oxygen minimum layer in the tropics and in upwelling areas can cause mass mortalities. Additional mass mortalities in coastal areas can be caused by 'red tides'.

Inadequacy of food may also cause increased mortalities. There are many other causes for mortalities, such as disease, spawning stress, and others which have been very little investigated in the past. The greatest cause for mortalities of fish, however, is predation (being eaten).

One of the environmental causes for mortalities of fish can be too high or too low temperature. The lowest and highest temperature at which a fish may survive depend upon its previous acclimatization. Therefore, sudden changes of temperature are normally much more dangerous to fish than slower changes during which they have time to become acclimatized.

Doudoroff's (1942) experiments showed that fish normally living in water of 12 to 25°C were killed by temperatures well above freezing point. He found that temperatures of 5° or even 10°C proved fatal to the species studied after acclimatization to 20°C. Mathews (1959) stated that mass mortalities can be caused by a change of as little as 5°C in less than a few days. Hayasaka (1934) reported that in the Pescadores Islands certain fish which live in shallow warm seas were killed in winter 1933-34, when the air temperature suddenly dropped below 10°C and stayed so for several days. Galloway (1941) recorded extensive mortality of fish in the coastal areas of Florida when the minimum temperature was as high as 14°C. From these cases it can be concluded that sudden drastic changes of temperature are lethal.

On the other hand, Simpson (1953) found that temperatures slightly under 0°C do not injure the cod nor the plaice of the North Sea. According to Simpson (*op cit*), W C Smith found that 43% of plaice survived in the hatchery when the water temperature was below 0°C. In his experiences at Espegrend, Leivistad (personal communication by Sjöblom) has shown that bottom fish (of the genus *Cottus*) survive at freezing temperatures.

Meuwis and Heuss (1957) studied the effect of temperature on the respiratory movements of carp. The results showed that the upper lethal temperature was lower for the older, heavier carp than for the younger, smaller fish. The younger fish showed little change in a temperature range of 15-35°C.

Blaxter (1960) found that for herring larvae 6 to 8mm long, acclimatized to temperatures between 7·5 and 15·5°C, the upper lethal temperature (defined as that temperature at which 50% of the fish die or become moribund after 24 hours) varies from 22 to 24°C and the lower temperature from —0·75 to —1·8°C.

Further description of mass mortalities of prawns, redfish, and cod caused by environmental conditions is given by Horsted and Smidt (1965).

Increased mortalities of fish can be caused by low oxygen content of the water. The distinct oxygen minimum layer is observed in tropical and sub-tropical parts of the oceans. This layer is usually better developed in the eastern than in the western parts of the oceans. The most pronounced oxygen minimum layer is found in the Arabian Sea, somewhat less pronounced in the Gulf of Guinea. In normal conditions this oxygen minimum layer is at 100 to 150m. The upper boundary of the layer is relatively sharp. In the core of this layer in the Arabian Sea there is no oxygen but hydrogen sulphide is present; in the Gulf of Guinea the corresponding oxygen content is about 0·5ml/l or some 10% of saturation. The Pacific oxygen minimum layer, appearing at greater depths, is relatively little developed. Although the oxygen minimum layer has been described as the layer of maximum consumption by some and as the layer of minimum replenishment by others, it cannot be said that either of the explanations is absolutely correct, since both factors probably are active in developing this more or less permanent oxygen minimum layer.

High consumption of oxygen close to the bottom occurs if the sediments contain much organic matter. This is usually the case over shallow shelves where upwelling normally occurs, causing high production of organic matter in the waters above. Oxygen can also be exhausted from more stagnant waters in isolated, poorly circulated basins of the shelf.

Upwelling of bottom waters from which the oxygen has been consumed by the organic sediments occurs usually in sub-tropical regions. Typical places for this kind of upwelling are the St. Helena Shelf and probably some areas in the Caribbean Sea (eg the Venezuela Shelf). In areas where upwelling occurs in sub-tropical or temperate regions over wide shallow shelves, there develops a high organic production in the water mass above the bottom. A great part of this organic production sinks to the bottom where it forms a sediment, rich in organic matter. Under normal upwelling conditions the water passes over the sediment rather quickly and the oxygen consumption in unit volume of water is relatively limited. Furthermore, the winds are usually strong and keep the water well mixed over the relatively shallow shelves. However, certain cessations of the usually more or less steady upwelling and mixing may occur. The weather is usually calm during these cessations and the mixing by wave action is therefore suppressed. Under these circumstances the water close to the bottom is exhausted of oxygen and becomes stagnant. The fish, which usually feed on the outer edges of the normal upwelling area, migrate towards the coast during the cessation of upwelling. When upwelling starts again, the bottom water now low in oxygen is brought to the surface relatively close to the shore, with the possible result of mass mortalities of fish.

Another cause of high mortalities of fish in limited areas close to the coast and in the estuaries, can be so-called 'red tides'. Red tides are caused by mass development of some plankton organisms, such as *Gymnodinium*, which can produce substances poisonous to fish.

One of the main causes of increased mortalities of fish might be inadequate supply of food. Unfortunately little is known quantitatively of the effects of starvation and starvation mortality, although it is generally assumed to be one of the main processes regulating the survival of larvae. It is often assumed, but seldom experimentally verified, that currents can transport larvae into the areas of either abundance or lack of proper food for larvae, thus affecting the survival of larvae. (Further information on fish and its food interactions is given in Chapter 4.)

Senescent mortality (mortality of old age) is assumed to be one of the causes of mortality in all fish. Recently it has been pointed out that spawning stress mortality may constitute a major part of the senescent mortality (Andersen and Ursin, 1977; Laevastu and Larkins, 1981).

Predation (*ie* being eaten by other fish and/or by marine mammals) is, however, one of the greatest sources of mortalities in juvenile fish. This aspect of mortalities and survival is emphasised in modern multispecies and ecosystem simulation models (*eg* Andersen and Ursin, 1977; Laevastu and Larkins, 1981).

The mortalities of fish, especially the 'natural mortalities' are still inadequately known. This can be summarized with the final statement from an extensive study by Jakobsson (1978) on the causes of the decline of Atlanto-Scandian herring: 'Taking into account the wide area of herring distribution in September 1969 it is suggested here that only a proportion of these herring found their way to the spawning grounds while the remainders have never been accounted for. They were never subjected to any fishing and must have suffered a very high 'natural' mortality because they could not be located in later years despite intensive surveying.'

3.3 Distribution of fish in relation to environment and its availability to fishery

Summary
The distributions of fish, specially pelagic species such as tuna, can be delineated by known optimum temperatures and monthly sea surface temperature charts. However, fish do not occur everywhere where its optimum temperature is found. Furthermore, in deeper layers, fish occur at different (usually colder) temperatures than indicated by sea surface temperature. Thus, in fish location by 'thermometric method' many other factors, such as season and seasonal migrations, occurrence of suitable food, *etc* are to be taken into consideration. Seasonal changes of temperature and occurrence of storms often affect coastal-offshore distribution of species.

Temperature and its changes may often be an indicator of other conditions and changes in the environment which might affect the distribution of the species directly and more than the temperature. Thermal boundaries (*ie* sharper horizontal temperature gradients) are often boundaries of surface currents, and affect the distribution of the species as well as their accumulation near these boundaries.

Fish search for and select a certain optimum combination of physical and biological conditions in the environment. Nearly all fish stocks have specific optimum temperatures. (Some optimum temperatures reported for different species and areas were summarized in *Table 2*.) A thorough knowledge of these optimum temperatures is necessary for the prediction of fish concentrations. With such knowledge, predictions of temperature, either statistical or synoptic, can be used for predicting the seasonal abundance of a given stock of fish to some extent.

The problem is further complicated by the fact that the environmental requirements change during the various stages of growth. Furthermore, some investigations indicate that the temperature requirements of a certain species also change seasonally (*cf* Devold,

1959), at least in connection with spawning. And, moreover, the concentration of food is temperature dependent, which makes the determination of the 'optimum temperature' for a fish complicated. In spite of this, the seasonal and year-to-year variations in thermal and other conditions will result in varying distribution and abundance on a given fishing ground. Actually, many fish make seasonal migrations towards the poles during the summer and towards the equator in winter. It may be that these migrations are either directly influenced by temperature or indirectly by the effect of temperature on the abundance of food. Apart from seasonal migrations, the shoaling connected with spawning, feeding, *etc* might be controlled by temperature, either directly or indirectly.

The best fishing grounds and areas are frequently located on the boundary regions of two currents or in other areas of upwelling and divergence. Uda (1936) stated that the saury (*Cololabis saira*) stays in the areas between the boundaries of the Kuroshio and Oyashio Currents, and is always migrating towards the maximum gradient of surface temperature. According to Uda (1952) the best fishing area for tuna is in the contact zone of these same currents (at temperatures of 18-20°C). The same applies to sardine, mackerel, and flying fish. Each of these species has a slightly different temperature preference. Similarly, Graham (1957) stated that in the central North Pacific the catches of surface-swimming and possibly also of deep-swimming albacore (*Thunnus germo*) were associated with the polar front and transition zone between central Pacific and subarctic waters. Surface catches were also associated with a seasonal latitudinal change in surface temperature, particularly about the isotherms 12·8–18·3°C. Kawana (1934) and Hida (1957) similarly reached the same conclusion that albacore tend to concentrate at certain surface isotherms (14–18°C, Kawana, *op cit*). Favourable catches of Japanese sardine on the southern grounds are made in the surface temperature range 12–16°C (Uda and Okamoto, 1936). Uda (1959a) has summed up the favourable temperature ranges for fishery of several important pelagic fish species in the NW Pacific. The temperature range for distribution and fishing of tuna species is shown in *Fig 15*.

The temperature ranges in *Fig 15* refer only to surface temperatures. Latest studies with albacore tuna, using sonic tags, show that albacore spends considerable time in thermocline layers where the temperature is 10 to 12°C (and dive even to 8°C waters, Laurs, pers. comm.), although the lowest temperature for albacore distribution in *Fig 15* is 14°C.

Yamanaka (1978) summarized the two existing theories pertaining to the distribution of tunas in relation to oceanographic features. The first is the 'current system theory' which tries to explain the relationship between the oceanographic conditions, in particular the pattern of ocean currents and the distribution of the tuna fished by the long-line fleets. In short it explains that 'Each species of tuna lives in a special current system according to its stage of ecology'. Thus the yellowfin in the Pacific Ocean lives in the area of the Equatorial Countercurrent and the South Equatorial Current; and the fishing areas are separated from those of the albacore which lives on the southern side of the Kuroshio Front and in the North Pacific Transition Zone.

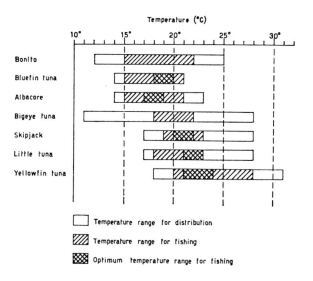

Fig 15 Temperature range for distribution and fishing
of tuna species (after Uda, 1952).

The second theory is from Uda (1973) who showed that good fishing grounds for
tuna are mainly found along the oceanic fronts around the equatorial upwelling and thus
in the area of both the Equatorial Countercurrent and the South Equatorial Current.
Thus so far as the yellowfin is concerned, the current system theory does not really
deviate from the upwelling theory, and the two major theories of tuna oceanography can
be considered as merged. The current system theory dealt mainly with the characteristics
of the meridional distribution but left the zonal one as largely uniform. Uda (1974) also
showed that there are concentrations of yellowfin along the Equatorial Countercurrent,
which correspond to the eddies along the convergence.

According to Yamanaka (1978), the knowledge of the general relationship between
the environment and tuna distribution are no longer useful, as the general distributions of
tunas have now been well mapped. What is now required is more detailed information on
the dynamic relationships between the changing oceanographic conditions and the
behaviour, distribution, and abundance of the tuna.

Dietrich, Sahrhage and Schubert (1959) found that in the northern North Sea the
concentration of herring may be influenced by four different phenomena in the distribution of temperature:

(*1*) In late summer and autumn the herring are concentrated in the core of the
cold bottom water. (The fishery workers of Japan and Norway have independently found out that the cold water pockets in the boundary regions have a
profound influence on the migrations of certain pelagic fish which tend to
aggregate in these pockets.)

40

(2) The lower the temperature of this cold water, the longer is the duration of herring concentration.

(3) The geographical position of this concentration fluctuates with the displacement of the centre of the cold water mass.

(4) The daily vertical movements of the herring schools are influenced (indirectly and directly) by the structure of the thermocline.

Rogalla and Sahrhage (1960) have described the occurrence of herring in the North Sea during June–July 1959 in relation to bottom water temperatures. Their principal findings are the following: In the central and northern North Sea cold bottom water spots can be found every summer. These are usually water masses which retain their winter temperatures because their mixing is prevented below the sharp thermocline. The herring seem to avoid those old cold waters (below 6·5°C) during the summer (*Fig 16*). Furthermore, they avoid the low salinity cold bottom waters flowing in the North Sea from Skagerrak. However, during the winter herring are found in the bottom waters of Skagerrak, which is then of Atlantic origin and has a temperature of 6-7°C. In the spring they move from this water into the feeding grounds in the North Sea.

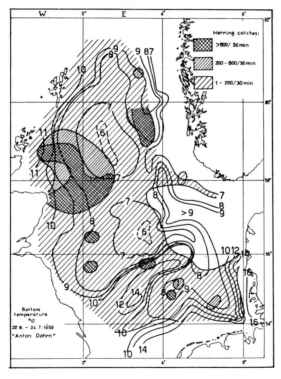

Fig 16 Occurrence of herring and the distribution of bottom temperature in the North Sea in June–July 1959 (after Rogalla and Sahrhage, 1960).

41

Jakobsson (1978) found that herring schools in the North Atlantic were usually found at the southern border of cold water fronts and in areas of mixed cold and warm waters. In June, herring occurred north of Iceland in waters with temperatures of 2 to 4°C, but some schools were found in waters less than 2°C. Later in the year most herring schools were found in 6 to 8°C waters. This illustrates that the temperature preference (if such exists) changes with the seasons. Capelin, however, was usually found in colder waters than herring, demonstrating the difference in temperature preference in different species.

Østvedt (1971) showed how temperature charts could be used successfully to locate herring at any time of the year because research on their behaviour related to water temperature had resulted in a good knowledge of the herring migration route throughout the year. Approaching the Norwegian coast the winter herring tended to gather in cold water pockets between Iceland and Norway before crossing, with increased swimming speed, the warm Atlantic current in order to penetrate into the colder Norwegian waters and to spawn there at temperatures between 6° and 8°C but avoiding areas with sea surface temperature below 5°C. Disappearing in May/June from coastal waters after spawning, the herring migrated to the boundary areas between the cold East Icelandic Current and warm Atlantic water masses. With this information on the migration route of the herring it might have been possible to establish a fisheries forecasting service which could have directed the offshore fishing fleet to the fish concentrations in the cold water pockets, and could have predicted with fairly high accuracy and considerable economic gains, the arrival of the herring in the spawning area. In addition, special cruises were jointly organized by various ICES member states in the Norwegian Sea which enabled the fisheries services of participating states to make pre-season forecasts.

However, Atlanto-Scandian herring stocks fell to a very low level in the late 1960s. Since 1965 the temperature on the north coast of Iceland had fallen below a threshold value for the herring and the herring migrated to more favourable conditions in the northern part of the Norwegian Sea. Although favourable conditions north of Iceland were restored later, the herring did not return. This example shows that sudden changes in environmental conditions may change the migration pattern of a fish stock for several years beyond the time when 'normal' conditions again exist. Fisheries forecasting services, therefore, cannot be based for the time being on definitely established relations, but need flexibility to make the best use of new results of research in this field.

The boundaries of the currents (convergences and divergences), and other dynamically caused oceanographic features (such as eddies), act not only as environmental distributional boundaries for fish, but cause aggregation of fish at these features by other properties. Zusser (1958) described a theory according to which the commercially significant aggregation of fish usually occurs in the centres of the current eddies where fish rest. His explanation points to the fact that fish usually stem the current but do need occasional rests.

Orton (1937) and Redfield (1939, 1941) have shown, at least partly, that the retention and accumulation of plankton, fish eggs, and fry occur in the centres of anti-cyclonic eddies. This aggregation might also be connected (*eg* through feeding) with the aggregation of adult fish in the current eddies. Local eddies are caused by the morphology of the coast and by the configuration of the bottom. Therefore, if the fish distribution is affected by eddies, knowledge of local details of currents could play an important role when looking for fishable concentrations of fish.

The current convergences bring about a 'mechanical' aggregation of forage organisms and small fish as well. This hypothetical mechanism of concentration has been described by Laevastu (1962). For a long time the Japanese tuna fishermen have made practical use of this knowledge to discover concentrations of pelagic tuna. The convergence lines between cold and warm currents are found to be especially rich. Here, in addition to the transport, the high basic organic production may play a significant role. Boundaries of currents may be either more or less permanent fronts at major currents or less marked changing lines at smaller current branches or at 'local' currents. In medium and high latitudes the convergences usually change their positions seasonally, causing seasonal shifts in occurrence of species.

It is well known to tuna fishermen that the best areas for bluefin tuna are connected with the maximum convergence and eddies of the frontal zones. Good albacore fishing is also usually connected with more or less local current boundaries or with the back eddies of oceanic islands and banks. The albacore occur also in cyclonically upwelled cool water masses along the warm-water side of polar fronts. These eddies and tongue-like pockets of warm or cold waters on meandering boundaries are areas of aggregation of other pelagic fish, such as herring in the North Atlantic and salmon in the North Pacific (*Fig 17*).

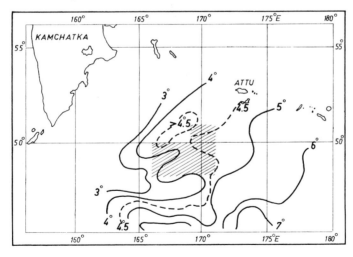

Fig 17 Japanese high-sea fishing area in the zone of tongue-like distribution of temperature (4° to 4·5°C), 10th July 1959 (after Terada, 1959).

43

The highest concentration of pelagic fish are found not only in the above 'tongues' and eddies, but also in the actual frontal zones, characterized by high gradients of various parameters, such as temperature, and thus by the narrowness of the zone, with optimum environmental conditions.

Analogously the aggregations of demersal fish are rather frequently determined by the current boundaries, the effectiveness of which in many cases depends on various local peculiarities, such as the sharpness and the stability in space and time of the boundary near the bottom, and availability of food on both sides of the boundary, *etc*. A sharp thermocline, intercepting the bottom, has an environmental effect very similar to that of a current boundary.

Magnusson *et al* (1979) studied the distribution of fish on the continental shelf of the Gulf Stream front just north of Cape Hatteras, where the front at the bottom was thermally very sharp indeed. Individual bottom trawl hauls spanned up to 8°C in less than 2km. The entire front spanned about 12°C. Thus, opposite sides of the front were close together in terms of space, but far apart in terms of temperature. The 12°C temperature difference is approximately one third the full range of temperature found in the sea and constitutes a long distance in temperature space. Similarities in species composition and the abundance of individual species in the trawl hauls suggested that organisms are distributed with respect to temperature space. Changes in the species structure of the fish community were most rapid immediately at the front (*Fig 18, upper*) and individual species change in abundance most rapidly at the front (*Fig 18, lower*). Distribution of southern species such as species (1) and (2) (*Fig 18, lower*) did not extend north of the front into cold water; northern species such as species (5) and (6) did not extend south of the front into warm water. A third response was exhibited by species [for example, species (3) and (4)] that reached maximum abundance in the intermediate temperatures of the front but did not extend either north or south of it. Outside of the frontal region the northern and southern species were broadly distributed over space. Their conclusion was that distance is best measured in a habitat dimension (*eg* temperature) rather than in geographic co-ordinates when we wish to locate particular species.

The behaviour of fish is not infrequently influenced by environmental factors other than temperature, but even in these cases the temperature may serve as a most useful indicator. This is illustrated by the observations of Davies (1956) who found that off the west coast of South Africa in general, but excluding Saldanha Bay, the maximum occurrence of juvenile pilchards (*Sardinops ocellata*) coincided with the minimum temperatures recorded during the summer months. It was also apparent that a relationship existed between the abundance of juvenile pilchard and wind. In spring and summer, when juvenile pilchard were most abundant, southerly winds were most prevalent. In autumn and winter, when southerly winds were far less frequent, juvenile pilchard were present in small numbers only. The seasonal variation in the abundance of juvenile pilchard also showed a direct relationship with the seasonal variation in the abundance of edible plankton.

It appears from the observations presented in this subchapter that many species tend to concentrate at certain isotherms and that current boundaries and areas with sharp horizontal temperature gradients form the limits to the distribution of certain species. Furthermore, aggregations of fish may occur at these limits as in the 'water pockets' close to them, as so clearly demonstrated by the Bergen school of fishery scientists. This fact was utilized in practice by the Norwegian as well as by the Russian herring fishery. As

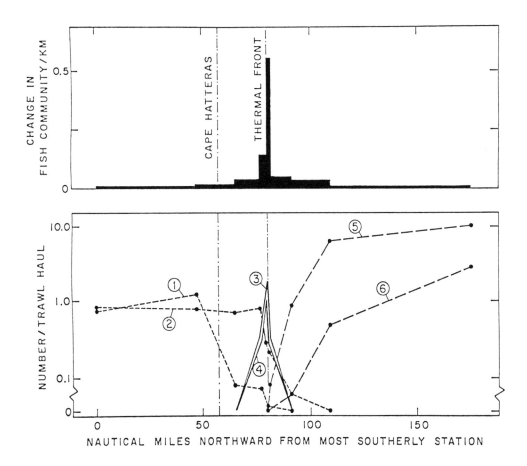

Fig 18 Changes in the distribution of fishes along the 30m depth contour on the continental shelf off North Carolina in the western Atlantic in May–June 1975 based on bottom trawl catches (from Magnusson *et al* 1979). Upper panel is the rate of change in species structure of the fish community based on the difference in similarity per kilometre between stations. Lower panel is the catch per trawl haul of six selected species that live south of the front (- - - -), in the front (————), and north of the front (— — —) between Gulf Stream and Virginian waters. The species are:
(1) offshore lizardfish, *Synodus poeyi;* (2) filefish, *Monacanthus* sp.; (3) croaker, *Micropogon undulatus;* (4) spot, *Leiostomus xanthurus;* (5) spotted hake, *Urophycis regius;* and, (6) Gulf Stream flounder, *Citharichthys arctifrons.*

another example it may be mentioned that the Japanese offshore salmon fishery in the North Pacific is confined to narrow temperature limits; fishermen and research vessels keep, in co-operation with the Japanese Meteorological Agency, a continuous check on the changes in sea surface temperature thus providing a basis for oceanographers on board the factory ships to make appropriate synopses and forecasts of sea temperature conditions, which are then communicated to the fishing fleet.

The temperature seems to affect also the coastal-offshore distribution of some fishes. Baxter (1967) indicated that adult anchovy become less available in inshore waters but are more numerous in offshore waters when water temperatures are warmer than average. According to Østvedt (1965) the time of migrations of Icelandic herring to the north coast seems to be affected by temperature, although the feeding migrations of herring are not fully controlled by temperature but are affected also by the availability of food (eg *Calanus*).

Cold bottom temperatures can affect the seasonal depth migrations of demersal fish. One example of such influence is schematically shown on *Fig 19*. Cold, sub-zero bottom waters are formed during some winters on the eastern Bering Sea shelf, which last until early summer. These sub-zero temperatures prevent flatfish (halibut, yellowfin sole, and others) from migrating to their traditional summer feeding grounds. Consequently aggregation of these fish occur at the boundaries of the cold water. No studies are as yet at hand on the effects of these delayed migrations on the survival and recruitment of the affected flatfish species.

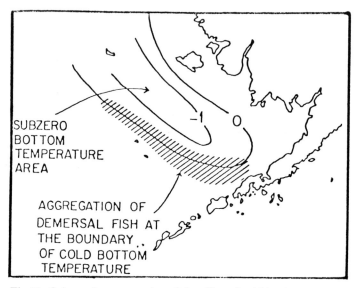

Fig 19 Schematic presentation of the effect of cold bottom temperatures on the distribution of demersal fish.

3.4 Long-term fluctuations in the abundance of fish and the effects of environment on these fluctuations

Summary

Long-term environmental changes could cause north-south shifts of spawning areas, cause changes in nursery areas, affect larval survival and growth rates, and change food availability. The long-term environmental fluctuations have small effects on truly demersal species. However, some changes in stock sizes and slight shifts of the distribution of pelagic and semipelagic species have been observed which might be related to long-term environmental changes. Most changes in the fish stock attributed to environmental causes have been related to exceptional conditions in one season (*eg* cold winters). Considerable changes in landings can be caused by the effects of weather anomalies on fishing conditions and availability to gear. There are considerable long-term fluctuations in all fish stocks which are mostly internal to the ecosystem and difficult to relate to external environmental influences.

Considerable interest has existed in the past in the possible climatic changes, especially in respect to long-term trends in temperature, mean surface wind, and precipitation. Many theories have been established on the sea-air interactions in the climatic scale. Lately these climatic studies have been intensified and correlations between climatic changes and changes of distribution and abundance of fish have been sought (further see Chapter 8 §8.3).

The long-term surface water temperature changes are not similar in all oceans, nor are they even similar in all parts of the same ocean. These changes are mostly determined by slight changes in transport in major currents and by the local meteorological anomalies.

Several workers have studied the correlation between long-term water temperature changes and changes in the distribution of species. In spite of the high degree of statistical correlation in some cases the causal explanation must, in general, be more complicated. Meyer and Kalle (1950) analysed the situation in 1938, when a close correlation could be shown between the long-term warming of the Baltic waters and the increase of cod catches. Their causal explanation can be summarized as follows: The warming of the Baltic waters was the result of an intensification of the atmospheric circulation which also resulted in the penetration of less diluted, more saline Kattegat water into the western and southern basins of the Baltic Sea. This penetrating water, being heavier than normal, displaced the stagnant bottom water of the basins and nutrients stored there, forcing it upward and mixing with near surface layers. This, in turn, led to a period of increased fertility over wide areas of the Baltic Sea. At the same time the deep waters were, of course, enriched with oxygen, which together with the rich phytoplankton, led to the heavy cod catches of the following years.

Lee (1956) concluded that the effect of climatic fluctuations upon the adult cod in the

Barents Sea is probably only slight. Täning (1953) summarized the probable ways in which a long-term temperature rise can influence the distribution of fish (in the Northern Hemisphere):

(1) spawning is diminished at its southern limit and increased at its northern limit,

(2) an increase in bottom water temperature can produce changes in spawning grounds,

(3) new nursery and feeding grounds become available in the north,

(4) an increase in the amount of food is brought about by the rise in temperature in higher latitudes and by changes occurring in currents and in the amount of nutrient salts present,

(5) the growth period is prolonged, and

(6) the limit at which larvae can survive is shifted further north.

Thus if, instead of the direct, apparent correlation, the more complicated but causally sound correlations are studied, in a few cases more significant factors can be found which explain the synchronism or time relation of long-term temperature trends with changes in fish distributions.

Several cases show that even the statistical correlation studies may give useful results: Rodewald (1955) discussed increased Greenland cod catches in the early fifties occurring simultaneously with the general increase of temperature in the North Atlantic, and believed that the small decrease of temperature in later years did not indicate a general return of the cold conditions prevalent before 1920. He also showed that the small catches of cod on the West Greenland Banks in July-August are related to the low temperatures of the Cape Farewell Current, which reaches the Banks during this season.

Uda and Okamoto (1936), observing the improvement in the sardine fishing in the Sea of Japan in the south during winter and spring, and in the north during summer, concluded that the fluctuation in the yield of sardine is related to the long-term changes in temperature. Uda (1952) concluded that the northward displacement of the main sardine spawning off Nagasaki was caused by the decrease of the coastal water temperature along the coast of Japan in preceding years; the fluctuation in the yield of sardine runs parallel to that of tuna, but inversely to the fluctuations of herring, squid, Pacific saury, and cod.

Grainger (1978) found that herring catches taken by Irish vessels off the Irish west coast for the preceding 80 years are probably reasonable indices of herring abundance in the area. The catches, which varied by two orders of magnitude, displayed a secular trend with superimposed short-term fluctuations. The secular trend appeared to be inversely related to climatic amelioration. The short-term fluctuations, dominated by a 10-year cycle, were related to sea surface temperature and salinity anomalies 3-4 years earlier, with winter temperature and salinity anomalies showing the strongest association. Oceanographic variation associated with the temperature and salinity fluctuations appears to affect these autumn-spawning herring in their first year of life, probably during the winter larval drift.

Hempel (1978) summarized the response of fish stocks in the North Sea to environmental changes with the following statements: 'It is generally accepted that climatic changes and fishing are the two most important factors influencing fish populations in the North Sea. Several attempts have been made to relate the year-class strengths of fish stocks directly to the wind (eg, Carruthers, 1938) and surface temperature (see review by Cushing and Dickson, 1976), but their direct effect on recruitment is rather limited. Nevertheless, these factors are important enough to influence the ecosystem in a way which finally results in responses by the stages of early life history. But the question about how the climatic changes affect recruitment remains. Poor year-classes are the rule and good year-classes are the exception in North Sea fish stocks and the building up of the stocks in the 1960s was due to a higher frequency of particularly rich year-classes. However, strong year-classes of closely related species do not necessarily occur in the same year (cod, haddock). No convincing argument has been developed to relate the recent cluster of exceptionally good year-classes in North Sea fish stocks to environmental factors. Did unfavourable factors, like predation, generally become reduced or did favourable conditions occur more frequently, both together resulting in higher magnitude and frequency of good year-classes? We know little about which factors or combination of factors favour or cause a strong year-class in a given species. Sometimes short events such as the temporary reversal of the current may have a considerable effect on the drift and survival of the larvae. Also the exact time for the determination of year-class strength is not well known, since there might not always be a short and well defined period.'

Hempel (1978) furthermore pointed out that the single stock concept, so prevalent in the past, is now being replaced by ecosystem approaches, taking into account the interactions between various fish stocks via their food and predation as well as the complex interaction with environment.

Holden (1978) studied the fluctuations of landings and fish stocks in the North Sea and found that there appears to have been a northward shift in the centre of population of several species, at least in the early sixties; in the case of plaice, turbot, brill, pilchard, anchovy, and horse mackerel this would have increased their availability in the North Sea and for haddock and whiting possibly decreased it. This northward shift of population might well be related to slight climatic change, eg increased flux of water to the North Sea through the English Channel.

Furthermore, Holden (1978) found that fluctuations in the landings of soles from the North Sea are closely linked with the occurrence of winters in which sea temperatures fall abnormally low. Such winters lead to high mortality of soles, some being killed directly by cold and others becoming so moribund that they are very easy to catch. In subsequent years abundance is low and catches fall. However, such winters are invariably followed by big year-classes which recruit to the fisheries 2 to 3 years later and which result in increased landings. Landings continue to increase until either another severe winter starts another cycle or the year-class is fished out.

Of interest is a summary (Sette, 1961) of the relations between the smallest and largest year-class strength. Sette found, for example, that among 21 successive year-classes of the eastern North Pacific sardine, the largest was 700 times greater than the smallest and in 28 successive year-classes of Alaskan herring the largest was only 34 times the size of the smallest.

Climatic changes are usually studied via changes of weather elements. Thus the studies of the effects of climatic changes on fish would be faciliated if there were direct relations between weather conditions and processes in the fish ecosystem. Nearly every fisherman is convinced about the existence of direct and indirect correlations between the climate, weather, and the fish. However, the true causes of correlations are not generally understood or known. It has often been attempted to correlate the recruitment of young specimens to a fish stock to weather conditions during the spawning season. A low water temperature (which is brought about by meteorological factors) can considerably delay the spawning or displace the spawning area so that the development of eggs and larvae is changed in time and place as described in Chapter 3 §3.1. This may result in low survival as the larvae grow up in areas where the environmental conditions, including the availability of food for them, are unfavourable. The availability of plankton as fish food is also to a large extent determined by the meteorological conditions, such as the amount of sunshine and the prevailing wind conditions, which determine the currents and the mixing of waters.

Hempel (1978a) has raised further hypothesis on the possible causes of changes in North Sea fish stocks, based on the changes in plankton in this area as ascertained by Glover *et al* (1974) (*Fig 20*). Hempel's hypothesis is reproduced verbatim below.

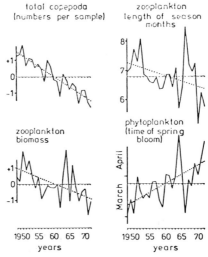

Fig 20   Changes in important parameters of plankton production in the North Sea. Data from continuous plankton recorder surveys (from Glover *et al*, 1974).

'In recent years several hypotheses have been developed regarding the causal link between recruitment and climate in the North Sea. Following Jones' (1973) argument on the importance of the particle size of food for the survival of fish larvae one might postulate that in recent years the composition of phytoplankton changed towards smaller size groups as suggested by the increased greenish colour of the continuous plankton records and a decreasing number of large diatoms and dinoflagellates in the 1960s. A decline in the abundance of large copepods, such as *Calanus*, and an increase in smaller ones, might have been a consequence of such changes in the food composition and might result in changes in the fish population, favouring fish which feed on small zooplankton. By a mechanism of this kind the composition of the fish population in the North Sea might be linked to cyclic changes of climate. We might also speculate that the decrease of *Calanus* in the northern North Sea in autumn had an indirect effect on the survival of autumn spawned herring larvae which live partly on *Calanus* nauplii. This seems a more likely cause of the decline in the herring stock than the drastic effect of the shortage of adult *Calanus* and *Spiratella* on the survival of adult herring.

'One would expect an increase in zooplankton, particularly *Calanus* after the reduction of its most important grazers, herring and mackerel. As this is obviously not the case, one might speculate that both kinds of fish also took great numbers of large carnivorous zooplankton, which now play the same role as herring and mackerel in earlier years by keeping the herbivorous copepods down.'

It is difficult to produce empirical proof to the above hypothesis as well as to many other similar hypotheses. There are long-period fluctuations in many fish stocks, *eg* in the Norwegian herring of thirty to fifty years, and in the Lofoten cod of three to four years, which might be somehow related to climatic fluctuations. Furthermore, the fluctuations in fish stocks can be caused by a multitude of factors and processes internal to the ecosystem, which often cannot be separated from external environmental influences (Laevastu and Larkins, 1981).

4

Fish, its food and environment interactions

Summary

Size and feeding regime-dependent (pelagic versus demersal) opportunistic feeding dominates in the marine fish ecosystem. Thus food composition of a given species varies considerably with age of fish as well as in space and time; past conventional food pyramid consideration is a gross oversimplification. Environmental factors especially affect the availability of pelagic food (plankton and nekton). Distribution and availability of food affects the distribution as well as the growth of fish. However, the distribution of pelagic food and fish is not always coincidental. Although herring are often found at the edges of large zooplankton patches, they are also found in areas of low zooplankton concentrations. Such occurrences might result from grazing or herring migrating in search of food. Migration routes of large pelagic fish, such as tuna and salmon, are assumed to be influenced by food availability.

The quantitative relation between basic organic production and the biomass of finfish is valid only in large regional scales. The best example of the effect of food availability on the stocks of pelagic fish is that of the Atlanto-Scandian herring north of Iceland where the low availability of zooplankton materially contributed to different distribution of herring and to the near-collapse of the herring stock in the second half of 1960s.

Good, comprehensive fish food studies are very few indeed. Thus there is an urgent need for such studies, specially in respect to the application of the results in trophodynamic computations in numerical ecosystem simulations.

A voluminous amount of literature is available on the food and feeding habits of fish. To summarize all that information is outside the scope of this book. Only a few salient points, in respect to the fish food, can be made here.

A schematic picture of food relations in the marine ecosystem is given in *Fig 21*. The actual food relations in the sea are, of course, still more complicated. The composition of food taken varies by age of the specimens, by location and by season, and is largely determined by the availability of food. Most fish species are opportunistic feeders. The

52

size of the food item in relation to the predator is a determining factor for the composition of food. The food cycle of cod in the Barents Sea is schematically shown in *Fig 22*.

The fish food in the sea can be classified as pelagic (plankton and nekton, both in the water mass) and benthic (on or in bottom living), bearing in mind that nekton (the organisms with ability to swim) includes fish and that 'big fish eat small fish', which is often overlooked in general trophodynamic (quantitative feeding) calculations. In some respects this division is arbitrary; however, it leads to a great simplification of the discussion. Some general relations between these different food types, the hydrographic conditions, and the fish are discussed in the following chapter.

It should also be noted that as the fish eggs and fry are important food items for many fish, the survival of a year-class often depends also on the density of predators.

The feeding behaviour of fish varies considerably from species to species. Hempel (1956) in his main observations on the feeding behaviour of the plaice and the lemon sole found that plaice were feeding during the day on bottom animals, especially on bivalves. During the night when the plaice were swimming in midwater, they were feeding on pelagic polychaetes (marine worms). While the plaice were feeding relatively little during the night, the lemon sole were feeding mainly during the night, eating more mobile animals (shrimp and other crustaceans) than the plaice. There is, even in the sea, often severe competition for food and living space between different species. The

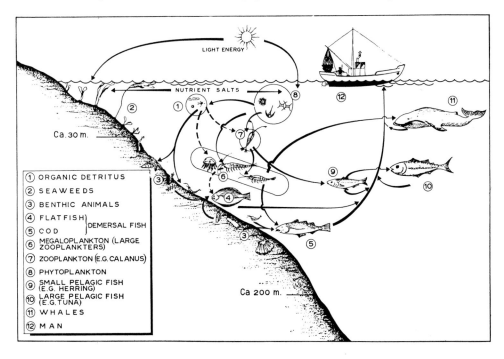

Fig 21   Food relations in marine ecosystem.

competition is largely influenced by changes of hydrographical conditions which may favour one and hinder other species in a given region.

Hempel (1978) in his analyses of long-term growth trends, found that for both pelagic and demersal species in the North Sea one phenomenon is common: an increase in growth rate, mainly observed in the first year's growth. Of the species considered in some detail, this holds true for herring, sprat, haddock, whiting, cod, sole, and plaice. The faster growth in length and weight led to earlier maturation. The faster growth would require also more food, and might be a consequence of increased food supply.

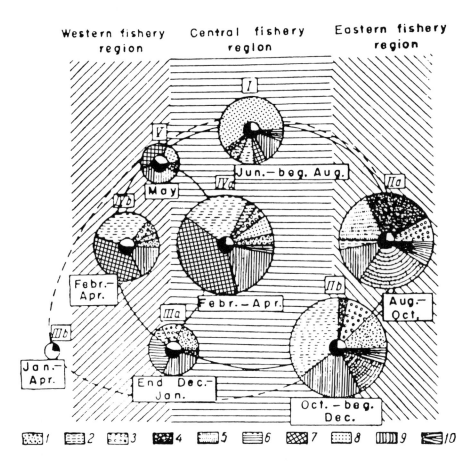

Fig 22 Scheme of food cycle of cod in Barents Sea (after Zatsepin and Petrov). Full lines, immature fish; dashed line, mature fish; I, summer food; II, autumn food; III, period of reduced feeding; IV, spring food; V, period of forced hunger. The area of circles indicates the index of fullness. 1, plankton organisms; 2, euphausiids; 3, shrimp; 4, benthic crustaceans; 5, other benthic animals; 6, herring; 7, capelin; 8, young cod, 9, polar cod; 10, other fish.

In plaice, the younger age groups grew faster from the fifties on, whereas in older plaice, growth was gradually reduced during the last two decades. In sole, growth in length increased steadily from 1963 in all age groups except the one-year-olds, an exception which is also true for plaice. Growth in the nurseries appears to be density-dependent, but there is no long-term trend in the available data. There is statistically significant evidence that the change in growth in sole is related to the development of the beam-trawl fishery which may have improved the availability of food.

Hempel (*op cit*) believes that timely availability of suitable food in the early life stages and intra- and inter-specific competitions may be the causes for the variation of growth as well as in determining the year-class strength. Andersen and Ursin (1977) found that food is normally not the major limiting factor in the productivity of North Sea fish stocks. On the other hand Steel (1974) estimated that most of the secondary and tertiary production must go into fish in order to sustain the present level of fish production.

Uda (1959c) stated that the migration routes of salmon (*Oncorhynchus*) in the North Pacific are determined by the line connecting the localities with abundant pelagic food, that is, areas which correspond to the eddies in the frontal zone of the Oyashio and the North Pacific Drift Current. These localities of abundant food vary from year to year in their time of development as well as in their north-southerly and east-westerly movements, which are related to the changing climatic conditions.

According to Yamanaka (1969) another explanation of the distribution of tuna fishing grounds is that the tuna is gathered by food organisms which are abundant in upwelling areas. This theory can be applied not only to the yellowfin in the eastern Pacific but also in the eastern Atlantic and east African waters. However, it does not fit the distribution of the bigeye tuna.

Many of the fish-food-environment interactions can be described via various trophic processes. These processes have been recently summarized quantitatively for the purpose of fish ecosystem simulation (Laevastu and Larkins, 1981).

The temperature of the environment can affect the food requirements of a given species and age via the temperature effects on metabolism and growth. The availability of proper food can also be affected by environmental factors (see the summary on Atlanto-Scandian herring at the end of this chapter).

The production and availability of pelagic and benthic food resources for fish affect the carrying capacity of different ocean regions in respect to fish biomasses. The fish carrying capacity is determined only in a large spatial scale (*eg* by the natural regions of the oceans), due to mobility of fish populations in search for food and suitable environment. The relations between fish biomasses and basic organic production is not a simple and linear one, as can be seen from *Table 4*. Many factors, such as depth, temperature (which affects the turnover rate), and type of biota present affect the relations between basic organic production and fish production.

The trophic (food) interactions are the most important interspecies interactions in

Table 4

General estimated biomasses of fish, benthos, zooplankton, and basic organic production

Type of area characteristics	*Total finfish biomass t/km²*	*Exploitable biomass t/km²*	*Sustainable annual yield t/km² (intensive fishery)*	*Basic organic production gC/m²/year*	*Zoo-plankton standing stocks t/km²*	*Benthos biomass t/km²*
*(1) Open continental shelves with 'upwelling' type circulation						
Tropics	25 to 45	8 to 15	3 to 7	} 200 to 400	50 to 180	20 to 100
Medium latitudes	40 to 60	12 to 20	4·5 to 8		} 100 to 300	50 to 200
**Higher latitudes	30 to 40	11 to 17	3·5 to 5·5			100 to 300
(2) Open continental shelves, no upwelling type circulation						
Tropics	15 to 30	4 to 10	1·5 to 4	} 100 to 250	30 to 120	20 to 100
Medium latitudes	25 to 45	8·5 to 15	3 to 5		} 80 to 200	50 to 150
Higher latitudes	20 to 35	8 to 14	2 to 4			80 to 200
(3) Wide marginal seas (eg North Sea)	25 to 45	9 to 18	6 to 7	100 to 200	80 to 250	100 to 400
(4) Semi-closed seas, Mediterranean type circulation	12 to 25	4 to 8	1·2 to 2·0	30 to 100	20 to 100	30 to 100
(5) Semi-closed seas, Baltic type circulation	18 to 28	5·5 to 9·5	2·2 to 3·5	50 to 150	50 to 150	50 to 100
(6) Open ocean						
Low latitudes	3 to 6	0·5 to 1·2	***(<0·3)	ca 50	30 to 150	—
High latitudes	5 to 12	1·5 to 3	***(<0·6)	ca 100	50 to 200	—

*In items 1 to 5 above the biomass and yield estimates refer to areas shallower than 500 metres.

** Assuming no great quantities of marine mammals present.

***These yields cannot be obtained due to dispersed nature of the resources.

the marine ecosystem. Environment can influence these interactions in a variety of ways, such as affecting the distribution patterns and therewith the predator-prey overlap and/ or separation in space and time. That predation, especially on larvae and juveniles, is a major component of mortality in fish biomasses can be demonstrated quantitatively by extensive stomach studies (*eg* Daan, 1973) or with large ecosystem models (Andersen and Ursin, 1977; Laevastu and Larkins, 1981). Cannibalism is also a common form of predation in marine fish ecosystem as a consequence of size-dependent feeding, but this and other forms of predation are affected very little by environment.

There are very few good, comprehensive fish food studies available in the literature which relate the fish food and feeding habits to the environment and anomalies in it. One of the recent and comprehensive studies in this category is that by Jakobsson (1978) on the Icelandic and Atlanto-Scandian herring. Jakobsson (1978) has described the distribution of Atlanto-Scandian herring in relation to zooplankton distribution, which is the main food for herring. Herring was usually found at the edges of higher zooplankton density areas and within higher zooplankton standing crop areas (*Fig 23*). Furthermore, the occurrence of herring was timed to 'biological seasons' – *ie* to the seasons of zooplankton development. However, some herring shoals were also located in low zooplankton density areas but in these cases the herring shoals were usually smaller, less dense, and migrated fast (*ca* 30n miles per day). Jakobsson concluded that the negative correlation sometimes observed between herring and zooplankton must mean that herring has grazed down zooplankton.

Jakobsson (*op cit*) found that the average adult herring had in its stomach 25ml of *Calanus* and made the following calculation on zooplankton-herring relation:

'In the low zooplankton density areas ($0–0.5$ml/m^3) each herring would have to eat every *Calanus* in 100m^3 to get 25ml. This corresponds to all the food under 200m^2 if the herring are mainly feeding in the uppermost 50m. A shoal of herring containing 0.5 million herring (about 120 tons) would have to eat every *Calanus* in an area of 100 million m^2 in order to fill their stomach once. The spring situation in the area north of Iceland is highly dynamic with regard to the "biological season" and the rapidly migrating herring. The herring and zooplankton distribution charts are, on the other hand, snapshots of the situation. The most common zooplankton herring relationship observed on the charts (*eg Fig 23*) is the one showing the herring at the edge of the zooplankton maxima.

'Taking these considerations into account it must be quite clear that the apparent negative zooplankton herring relationship appearing during some surveys is either because the herring have grazed down or are in the process of grazing down (and this is the most common situation) the dense zooplankton concentrations. One also must not forget that sometimes the herring were passing through low density zooplankton areas on their way to or while searching for better feeding conditions.'

A few additional excerpts from Jakobsson (1978) study describe the collapse of north Icelandic herring stock and causes of this collapse:

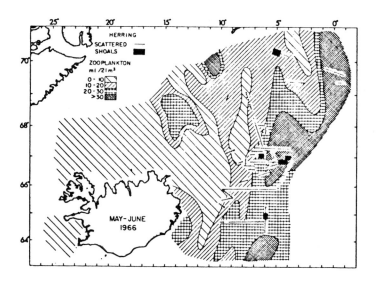

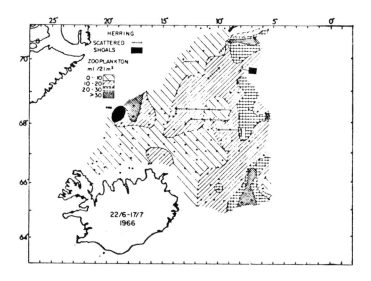

Fig 23 Distribution of zooplankton and herring NE of Iceland in
May to July 1966 (Jakobsson, 1978).

'In 1965, 1967, and 1968 most of the traditional offshore herring grounds, which were ice-free in earlier years, were blocked by ice in June-July while polar water covered the ice-free areas. The temperature in the East Icelandic Current fell drastically and large areas were covered with water of —1·8°–0°C. During these ice-years 1965, 1967, 1968, and 1969 the average primary production off the western and middle north coast was only about a quarter of that observed in the period 1958-1964 while such drastic reduction did not take place at the south and west coast of Iceland. It is probable that the greatly reduced primary production caused the observed complete collapse of the main stocks of zooplankton in the Iceland north coast area.

'After 1970 the primary production returned gradually to former high levels while the concentrations of zooplankton remained at very low levels. The evidence therefore indicates that the adverse conditions north of Iceland must have practically destroyed the *Calanus finmarchicus* population since it has taken so long to recover. It also follows that a major part of the *Calanus finmarchicus* population has been a local stock of the north Icelandic waters rather than one brought there every spring by the warm Atlantic current.'

There can be no doubt that the drastic environmental changes which took place north and northeast of Iceland in the mid-sixties and described by Jakobsson made radical changes in the feeding migrations of the Norwegian herring inevitable. Thus changes in the environmental factors were responsible for the collapse of the fishery in the traditional north Icelandic waters.

5

Migrations of fish and their relation to environment

The migrations of fish have been summarized in detail by Harden Jones (1968). In this chapter we will review only a few aspects of fish migrations in relation to environment, emphasizing those aspects of fish migrations and movements which have received less attention by Harden Jones.

The migrations of fish and their causes are schematically shown in *Fig 24*. Harden Jones (1968) suggests that fish movements can be summarized in the form of triangular patterns in which adult fish move between their spawning, feeding, and wintering areas in a sequence which depends on the season in which they spawn: spring spawners follow a clockwise sequence of feeding-wintering-spawning, and autumn spawners a counter-clockwise sequence of feeding-spawning-wintering. Such a simple scheme does not always fit all known migration patterns and every species and region must be considered a special case.

Plausible ranges and time scales of advection (passive transport) and migrations are shown schematically in *Fig 25*. The average advection (about 1km/hr) and average

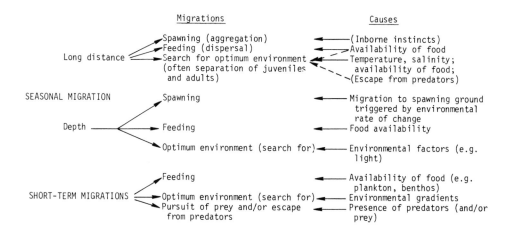

Fig 24   Scheme of environmental causes for migrations.

migration (about 3km/hr) are indicated relating time to distance and periodic migrations such as semidiurnal and seasonal migration are also shown. The expected relative magnitudes of changes of biomass of a given species as a result of different periodic migrations as well as possible long-period 'drift' within the system (which could result in long-period changes in abundance depending on excursions of pelagic fish on and off the shelf and demersal fish into and out of the areas along the shelf) are shown in *Fig 26*.

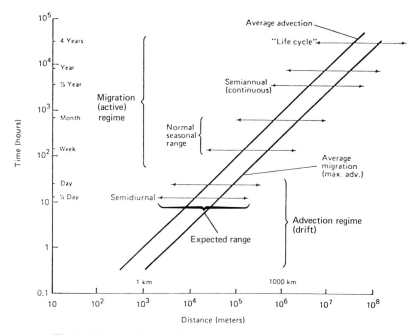

Fig 25   Ranges of advection and migration of stocks in marine ecosystems.

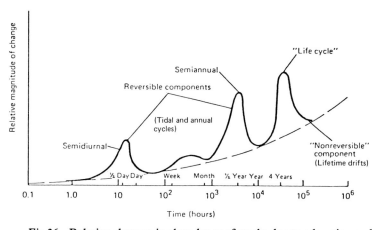

Fig 26   Relative changes in abundance of stocks due to advection and migration.

5.1 Influence of environment on shoaling and dispersal, and swimming speeds of fish.

Summary

Two main reasons for shoaling are considered to be: the minimization of predation and enhancement of spawning success. Shoaling is affected by light, both seasonally and diurnally. Most fish usually disperse during the night.

Many smaller pelagic fish occur in small, scattered shoals during feeding, but form large shoals during spawning migrations and spawning. Overwintering shoals tend also to be large. Mixed shoals of different species but of the same size fish, have been observed.

The swimming and cruising speeds of fish are a function of fish size. Fish cannot maintain maximum (escape) swimming speeds for any considerable length of time. The speed of fish is also affected by temperature.

The factors affecting shoaling and dispersal and the size of shoals are not well known.

The shoaling behaviour and shoal sizes and patterns vary considerably from species to species and also in space and time. The shoaling varies also throughout the life span of the species. It is generally assumed that shoaling has two main functions – to minimize predation and to enhance spawning. There is also aggregation on the wintering grounds, and most fish undertake migrations in shoals.

Many pelagic fish aggregate on the patches of food (zooplankton) and disperse in small shoals when the food patch has been grazed down. Jakobsson (1978) in his extensive studies of Icelandic and Atlanto-Scandian herring found that when herring food is low, the fish is dispersed. The feeding shoals are in general small and scattered, whereas spawning shoals are large, as are overwintering shoals of some species.

Horwood and Cushing (1978) pointed out the seasonal difference in shoaling in mackerel. This species is found in European waters during the summer in small shoals, while the fish is feeding. During the winter on the overwintering grounds, mackerel forms large shoals during the day, while it disperses during the night. Horwood and Cushing (*op cit*) gave two examples of the sizes of overwintering shoals: one had the dimensions of 5×1.5n miles and was 12m thick, containing 750 million fish, the other was 24×7n miles and contained 990 million fish.

Shoal sizes vary considerably; those reported are usually the larger and more spectacular shoals. Jakobsson (1978) reported that normally the Atlanto-Scandian herring occurs either in large, plume-shape shoals of 50 to 500 tonnes, or in small clusters of 10 tonnes and less.

The shoal can be mixed, containing different species of about the same size fish (Radovich, pers. comm.) .Furthermore, the shoaling behaviour of the fish can be dependent on the stock size and density. Jakobsson (1978) reported that in the second half of the 1960s, when Atlanto-Scandian herring stocks started to decrease, it was found that herring shoals contained a mixture of herring, blue whiting and sometimes capelin.

Shoaling and dispersal of shoals are greatly affected by changes in light intensity (Shaw, 1961). The minimum light intensities, at which shoals disperse, differ widely for different species. However, the behaviour of fish in respect to light also depends upon the turbidity of water. The behaviour is also affected by other environmental factors, *eg* by temperatures, which usually are significant to the physiology of fish, as well as by the spawning habits. Little information is available on the diurnal cycle of shoaling tendency in respect to light. Schärfe (1951) found from echo soundings in Plön Lake that, at the time of observation, fish had a tendency to gather in small shoals during the day but dispersed during the night. Hunter (1968) demonstrated that fish shoals disperse during dark nights and that no food is taken in darkness. However, the threshold light value for dispersal was found to be relatively low. Thus during nights with moonlight some species (*eg* jack mackerel) might not disperse and will be able to feed near the surface.

Blaxter *et al* (1958) found that there is a critical light intensity for shoaling and obstacle avoidance by adult herring. Shoaling began to cease at about 0·1 lux in fish adapted to light, and at 0·001 lux in fish adapted to darkness. Obstacle avoidance became ineffective at about the same intensity by light-adapted fish and at a lower figure by darkness-adapted fish. Further notes on shoaling behaviour in relation to diurnal migrations are found in the next subchapter.

The migration speeds of fish shoals vary widely. Examples of estimated migration speeds are given in *Table 5*. In addition, Jakobsson (1978) found that Atlanto-Scandian herring shoals migrate about 15n miles per day. However, migration speeds in excess of

Table 5
Speeds of migrating fish

Species	Speed (km/d)
Sole[1]	7–16
Plaice[1]	1–7
Herring[1]	4–30
Salmon (sockeye)[2]	54
Salmon (chum)[2]	48
Halibut[3]	6
Herring[4]	25
Yellowfin sole[4]	3–7

[1]Harden Jones (1968) – North Sea area
[2]Kondo *et al* (1965) – East Bering Sea shelf
[3]Novikov (1970) – East Bering Sea shelf
[4]Estimated from spawning migration – East Bering Sea shelf

1·4 knots were also observed. In low zooplankton density areas, the herring shoals moved about 30n miles per day.

The swimming and cruising speeds of individual species are higher than the mean migration speeds. Some data on the observed swimming and cruising speeds of various fish have been summarized in *Table 6*. Obviously the swimming and cruising speeds must vary with the size of fish and also from species to species. Some observations also show that the swimming speed is affected by the prevailing temperature. Brett *et al* (1958) found the maximum cruising speed of a sockeye salmon to be at 15°C and that of coho to be at 20°C (see *Fig 27*). However, Blaxter and Dickson (1959) were not able to find any obvious correlation between the average maximum swimming speed and the temperature. These authors estimated that small fish (smaller than 30cm in length) can swim about ten body lengths per second; the maximum speed of larger fish is relatively less.

It has been observed that fish cannot maintain their maximum speed for a long time and that their cruising speeds are usually much lower than the maximum swimming speeds observed in an experiment. Some idea of the cruising speeds of fish can be obtained also from pelagic trawling experiments. Schärfe (1959) found that the minimum speed required to catch sprat with a pelagic trawl is 2·5 knots and to catch bank herring in the North Sea 3·5 knots. From various data it can be tentatively concluded that for pelagic fish of small and medium size the cruising speed is *ca* 2·5 knots during the day-time while the day and night average is *ca* 1·5 knots.

Table 6
Examples of swimming speeds of fish

Species	Speed (cm/sec)	Remarks	Authors
Young sockeye and coho salmon	30–40	Maximum sustained speed	Brett, Hollands and Alderdice (1958)
Herring larvae	0·6–1·0	Critical velocity	Bishai (1960)
Herring	91	Maximum speed for 15·2 cm fish	Brawn (1960)
Herring	143	Maximum speed for 26·7 cm fish	Brawn (1960)
Cod (12–56cm)	75–210	Maximum speed	Blaxter and Dickson (1959)
Herring (1–25cm)	3–170	Maximum speed	,,
Mackerel (33–38cm)	189–300	Maximum speed	,,
Plaice (6–25cm)	6–129	Maximum speed	,,
Horse mackerel	5–55	Observed speeds; most frequent 25 cm/sec	Kawada, Tawara and Yoshimita (1958)

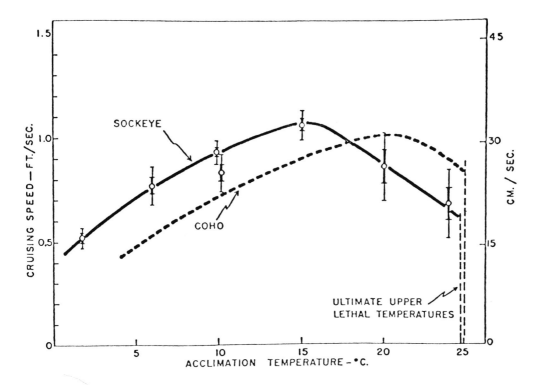

Fig 27 Variations in cruising speed for temperatures acclimated underyearling sockeye and coho adjusted in each case to common mean lengths of 6·9cm and 5·4cm respectively. The samples were cultured under similar conditions and are of comparable age, four to six months from hatching (from Brett *et al*, 1958).

5.2 Fish distribution with depth and vertical migrations

Summary
The diurnal vertical migrations have been systematized into six different categories. Most of the vertical migrations are affected by diurnal change of light intensity. However, the thermal structure (specially thermocline) and wave action also influence the depth distribution of fish. Trawl catches vary diurnally depending on the diurnal behaviour patterns of the species.

Vertical migrations of demersal fish near the bottom can also occur in tidal rhythm with the result that certain fish utilize tidal currents for enhancement of migrations.

The vertical migrations of fish vary diurnally and seasonally, thus observations of fish distribution with depth must take the time of the day and season into consideration. Marine animals can be divided into the following six groups by the nature of their diurnal vertical migrations (*see Fig 28*).

(A) Pelagic species with daytime occurrence slightly above the thermocline; migration to surface layer at sunset; dispersion between surface and thermocline during the night; descent to above the thermocline by sunrise.

(B) Pelagic species with daytime occurrence in layers below the thermocline; migration through thermocline into surface layers during sunset; dispersion between surface and bottom during the night with bulk occurring above thermocline; descent through thermocline into deeper layers during sunrise.

(C) Pelagic species with daytime occurrence in layers below thermocline; migration to thermocline during sunset; dispersion between thermocline and bottom during the night; descent into deeper layers during sunrise.

(D) Demersal species with daytime occurrence on or close to the bottom; migration and dispersal into the water mass below (and occasionally also above) the thermocline during sunset; descent to the bottom during sunrise.

(E) Species which are dispersed throughout the water column during the day but which descend to the bottom during the night.

(F) Pelagic and demersal species without any distinct diurnal migrations.

In general, most pelagic fish rise to the surface layers before sunset, usually in shoals. After sunset they disperse in the water column, and sink into the deeper layers by sunrise

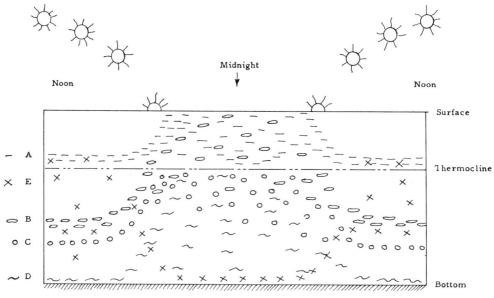

Fig 28   Schematic presentation of five different types of diurnal vertical migrations.

66

(Groups A, B, and C). Demersal fish usually spend the day on the bottom but rise and disperse in the water column during the night (Group D).

The effectiveness of fishing with lights before midnight indicates a maximum of phototaxis of some fish at that time. This may also be due to the fact that some fish are in general more active in the evening and before midnight.

According to Nomura (1958) sardine (*Sardinops melanosticta*), and probably many other species, are most active during the twilight of morning and evening. During the day they descend into waters deeper than 30m, at an illumination of 10-1 000 lux, but rise to the surface during the twilight and descend again at night (Group A). In grey, cloudy weather the behaviour of sardine in respect to light is not very pronounced.

Richardson (1952) found that off the Thames estuary sprat started to rise towards the surface about one hour before sunset, and after sunset the rise was considerably accelerated. They left the surface by sunrise. Pilchards were attracted to lights during the night. The behaviour of herring varied slightly in different areas in the North Sea. The depth level of herring shoals was found to be determined by the intensity of light. Adult herring are not very positively phototactic and prefer lower light intensities. Herring are attracted to artificial light at night if it is not too bright, but keep a considerable distance from strong light.

Postuma (1957) found that the depth of herring shoals at night depends on the depth of the mixed, homothermal surface layer. That water temperature was a controlling factor was shown by the fact that herring did not penetrate far into the mixed surface layer. Thus his observations clearly demonstrated that at least two factors are involved in the pattern of diurnal migration of herring: light and temperature (Group C).

Radakov and Soloviev (1959) observed from a submarine that herring grew more active in the evening and throughout the night, reaching the peak of activity at dawn. The herring moved upwards in the evening but from midnight onwards, and especially during the day, they showed negative phototaxis and were relatively passive during the day. This observation also explains why Brawn (1960) did not find any correlation between the mean solar radiation for daylight hours and the median depth of herring.

Some examples of the diurnal migrations of demersal fish are also known from literature. Ellis (1956) found that cod in the North Sea was in small, tight shoals during the day, about 0 to 18m from the bottom. During the night they dispersed in a thicker layer 0 to 55m from the bottom, the total water depth being 100m (Group D).

According to Baxter (1967), anchovy form, during the day, bands of shoals on the bottom and rise during the night towards surface where they form bands 6 to 15m thick (Group D).

Schmidt (1958) stated that species of adult coalfish (*Pollachius virens*) are active in the water during the night but sink to the bottom in the morning and remain there during the day (Group D). These big fish usually live on flat offshore banks. Younger specimens of the same species live on the slopes of banks and behave differently.

Harder and Hempel (1954) found that, with normal diurnal light conditions, plaice were active only during the hours of darkness. If the illumination in the experimental tanks was not switched off for the night time, normal nightly swimming activity was suppressed, but the plaice were abnormally active during the following day.

Russell (1928) investigated the diurnal up and down movement of fish larvae and young fish in the Plymouth area and found that only young clupeids and gobies showed marked vertical diurnal migrations.

Woodhead (1965) concluded that many demersal species spend a considerable amount of time in midwater, which fact reduces the catch-rate of bottom trawls. Woodhead gave detailed observations of diurnal behaviour of many commercially important fish species in the North Atlantic.

The depth of the pelagic shoals depends largely on the vertical temperature structure. It is known, for example, that some pelagic fish swim deeper when the surface waters are warm. Postuma (1957) has indicated that the depth of herring shoals at night depends on the depth of the mixed surface layer, since the herring do not penetrate far into this layer. The pattern of the diurnal vertical migration of herring must be affected, both directly and indirectly, through the concentration of food at the thermocline. Similarly, Berzins (1949) has shown that the pelagic fish in the Gulf of Riga (at night) avoid extremely warm and extremely cold water. In July and August most sprat (*Sprattus sprattus balticus*) are caught in the temperature range 10-15°C, while most of the small Baltic herring are caught at temperatures of 8-12°C.

There are several aspects of fish behaviour (*eg* feeding response to current, shoaling) which change diurnally and may be directly or indirectly influenced by light. Also, in certain cases, the reactions of fish in respect to other marine ecological groups may be influenced by light conditions. One of these examples is the relation between fish and phytoplankton.

As mentioned earlier, some phytoplankton organisms are able to produce during the photosynthetic period toxic substances which cause the fish to avoid concentrations of phytoplankton during the daytime. Therefore, the diurnal up and down migrations of fish in respect to light may also be influenced by phytoplankton. The standing crop and the photosynthetic activity of phytoplankton show large seasonal fluctuations. If the phytoplankton plays an important role in the vertical migrations of fish, these vertical migrations can be expected to have the same kind of seasonal variations as the phytoplankton.

Inoue and Ogura (1958) found that the depth where anchovy shoals swim varies during the day, being shallow in the morning and evening. This depth also varies according to different weather conditions and between different fishing grounds, but does not vary with size of fish and season. These observations have a natural explanation, as anchovy prefer a certain light intensity, and the depth of the light intensity varies with the hour of the day and according to cloudiness. Furthermore, this depth of a given light intensity is determined also by the extinction coefficient of the water (*ie* turbidity) which can vary from one fishing ground to another.

The vertical migrations of flatfish might be triggered by and synchronized with tidal currents, as observations of plaice, tagged with acoustic transponding tags, indicate. An excerpt of the report on this subject by Greer Walker *et al* (1978) illustrates the above statement:

'The most consistent, and the most interesting, behaviour pattern observed in the tracking work is that which we have called selective tidal stream transport: fish leave the bottom at slack water to move downstream in midwater on one tide, and return to the bottom at the next slack, with little, or no significant movement on the opposing tide. Plaice behave in this way both by day and by night.'

The diurnal vertical migrations affect the trawl catches considerably. Parrish *et al* (1964) have summarized the vertical migrations and related them to trawl catches as follows:

'Horizontal distribution of the fish often varies with the age of the fish and its physiological condition and, therefore, with the season, and also with such environmental factors as food, temperature, and salinity. Vertical distribution often changes with these factors also, but it is its diurnal variation in relation to changes in light intensity which is often the most dominant factor affecting catches with a particular gear; this is well known amongst the clupeid species (*eg* herring). Similarly, the behaviour of fish in relation to the gear may change with size, age, and physiological condition, but the principal factors governing it are the types and strength of stimuli produced by the gear. Of these, visual ones are certainly of major importance, and so are also subject to marked diurnal changes.

'Diurnal changes in light thus have an important influence on at least two major factors which may give bias in catch data.

'Diurnal variations in trawl catches is very marked for clupeid species, especially herring, whose vertical distribution changes from the deeper water layers by day, where they are fished by bottom trawls, to the upper water layers by night, where they become available to drift-nets, pelagic trawls, and encircling gears (purse-seines, ring-nets). Although the diurnal variations in vertical distribution of most demersal species are generally less marked, diurnal changes in the catches of a number of them are well-known to fishermen in many areas.

'Diurnal variations in catches may differ in both direction and magnitude because of the relative importance of different causal factors; in some instances the effects of these factors, operating in opposite directions may counterbalance one another, thereby resulting in no substantial differences in catches, or they may result in a reversal of the difference between species, areas, and times. In those fisheries where the daylight catches are significantly greater than the darkness ones, it seems likely that the principal factor is the diurnal change in vertical distribution, probably in relation to feeding habits, which for any species may differ between localities and seasons. However, where the darkness catches are greater than the daylight ones, it seems that the main cause must lie in the differences in visual responses of the fish to the gear.'

5.3 Spawning, feeding, and other seasonal migrations

Summary
Most of the seasonal migrations are into shallower water during the spring and into deeper water during the autumn. These migrations might be both temperature and light intensity controlled. Therefore, environmental anomalies affect the timing of these migrations. Spawning and feeding migrations are species specific.

The quantitative aspects of most seasonal and life-cycle migrations are poorly known for most species and areas (with possible exception of the North Atlantic).

Beside the diurnal up and down movement, most species undertake seasonal vertical migrations (see *Fig 29*). These seasonal vertical migrations are, in some species, *eg* cod, partly governed by light conditions (Trout, 1957).

Both pelagic and demersal fish undertake horizontal seasonal migrations, usually into shallow water or closer to the surface during the summer and into deep waters during the winter. As possible causative factors and 'triggering mechanisms' can be

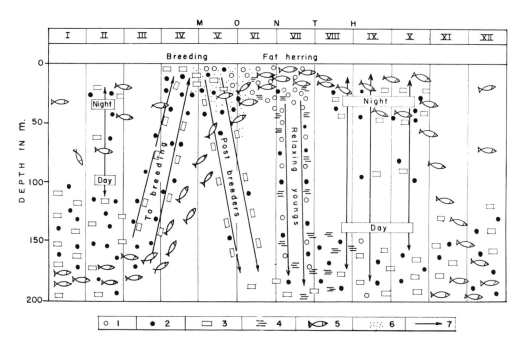

Fig 29 Scheme of the seasonal migrations of herring and its food (Calanus and Euphausiids) on Murman Banks (after Manteufel). 1, *Calanus finmarchicus*, stages V–VII; 2, *Calanus finmarchicus*, stages I–IV; 3, adult *Thysanoessa*; 4, larval *Thysanoessa;* 5, herring; 6, high phytoplankton concentration; 7, migrations.

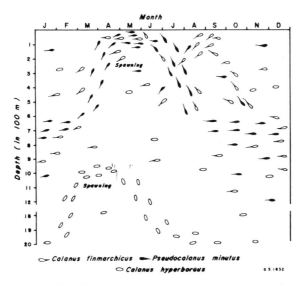

Fig 30   Vertical seasonal spawning migrations of
some zooplankton organisms in the Norwegian Sea
(data from Østvedt, 1965).

listed temperature, light and seasonal migrations of the available food. *Figure 30* shows
the great vertical seasonal spawning migrations of some zooplankton organisms in the
Norwegian Sea. Such seasonal vertical migrations of fish food organisms may cause the
fish to follow the food abundance.

The seasonal tendencies to up and down movements of demersal fish result in
shoreward and seaward migrations. Knowledge of these bathymetric movements is
important in fishery management, since they explain the dynamics of the movements of
exploitable demersal fish. Alverson (1960) found off the northwestern coast of North
America bathymetric movements for English sole (*Parophrys vetulus*), petrale sole
(*Eopsetta jordani*), Dover sole (*Microstomus pacificus*), starry flounder (*Platichthys
stellatus*), cod, ocean perch (*Sebastes marinus*) and sablefish (*Anoplopoma fimbria*).
Lingcod (*Ophiodon elongatus*) appeared to migrate from areas normally fished by
trawlers during summer months. It was presumed that this was an emigration of mature
fish to spawning areas not accessible to the fishing fleet. The apparent seasonal vertical
migrations result in seasonal changes in the productivity of fisheries and in the composition of catches. Highest individual catches for some species are made during the winter
months, but highest total catches are taken during the summer. Trout (1957) investigated
the seasonal migrations of Bear Island cod, utilizing the information on commercial
catches. His mean annual depth curve is shown as *Fig 31*. (The broken line in the figure
is an extrapolation based upon the hypothesis that the immature cod, when not fished,
overwinter in deep water of the West Spitsbergen Current.) He concluded: 'Water
movement is responsible for the changes in distribution of the Arcto-Norwegian cod,

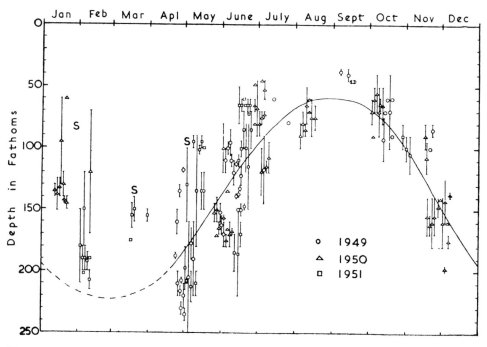

Fig 31 Mean annual depth curve of cod, Spitsbergen Shelf west of 21°E, from commercial trawler W/T reports received by *Ernest Holt*, 1949-51 (after Trout, 1957).

yet this is only a secondary effect of their changing behaviour in response to changing light conditions, which results in their primary annual vertical migrations and their annual depth range. During the summer, when cod schools are largely pelagic, they are displaced horizontally by currents. Winter migration takes place on the bottom in the absence of light. Normally the fish will tend to remain in their particular water mass. Nevertheless, a transfer from one water mass into another may take place as a result of changing behaviour requirements. In general, in the absence of water flow, the horizontal movements of Bear Island cod would be of limited extent; however, the annual vertical migrations would be expected to persist.'

The seasonal, temperature controlled, onshore-offshore migrations occur in many species. Such migrations of Japanese sardine are illustrated schematically in *Fig 32*. Bitiukov (1959) cites the work of Pirozhnikov who found that the periodic shoreward migrations of the muksun (*Coregonus muksun*) in Tiksi Bay were controlled by the oxygen content of water. The coastal water was rich in oxygen but poor in food. The deeper offshore water contained plenty of food but little oxygen, and the fish could not remain in this water for longer periods. Similar oxygen controlled sporadic mass shoreward migrations of demersal fish and crustaceans in Mobile Bay, Alabama, were reported by Loesch (1960): 'The Mobile Bay jubilee, in which crabs, shrimp, some fish, and other estuarine types crowd to the edge of the water at infrequent intervals during

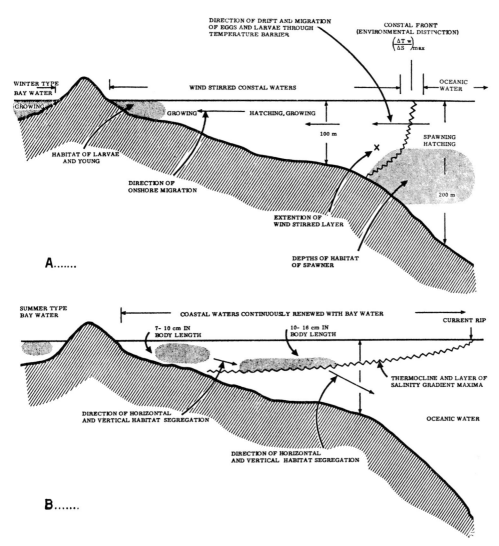

Fig 32 Schematic representation of the offshore-onshore movements of the Japanese sardine (*Sardinops melanosticta*) in the coastal waters of western Japan (according to Tsujita, 1957). (A . . . Winter season; B . . . Summer season.)

summer months chiefly along the northeast shore, occurs most frequently during the dark hours on a rising tide following a day of east wind. The coincidence of these climatological conditions results in an inshore movement of water which is low in oxygen because of the rotting of wood debris accumulated from the Tensaw River run-off from a deep pocket near the eastern shore of the Bay. Demersal animals especially move inshore ahead of this water mass. Changes in tide, wind, or other meteorological conditions result in water aeration and terminate the jubilee. Mortalities are rare.'

73

Galtsoff (1924) and Pektas (1954) have related the seasonal migrations of mackerel in the Black Sea to the temperature. The seasonal, temperature-controlled migrations of mackerel in the North Sea were described by Dannevig (1955), who pointed out the fact already known to fishermen, that the mackerel in north European waters disappear from the surface when the water gets colder, and appear again in the spring when the water starts to warm up. In the higher latitudes the mackerel stay at a depth of several hundred metres during the winter. The migration to the surface starts in the spring when the bottom water grows colder. After a long severe winter the mackerel therefore migrate to the surface earlier than after a mild winter, when the temperature of the bottom water is relatively high. Mackerel seem to avoid water which has a temperature lower than 4-5°C. Similarly, Jackman and Steven (1955) correlated the arrival and departure of the mackerel to the sea surface temperature at Torbay, and also to the establishment and breakdown of the summer thermocline.

The seasonal behaviour and migrations of fish might also be related to current patterns in the following way. The eggs and larvae are transported with the current away from the spawning ground. At a certain stage of life the fish begin to swim actively against the current to reach the spawning grounds again. This migration cycle may even be an annual one. The spawned fish, being weak after this process, might be carried away with the currents to the feeding grounds. Later, they must swim back to the spawning grounds against the current, thus completing one cycle per year. This can be illustrated with the example of the Arcto-Norwegian cod which range over the whole of the Barents Sea and West Spitsbergen waters in their summer feeding migrations and come together in the early spring to spawn off the Lofoten Islands. The current charts show that fish and fry as well can be carried in the West Spitsbergen and North Cape Currents up to the Bear Island area. The products of the spring spawning in the West Fjord are thus distributed in autumn over the feeding grounds of the western Barents Sea. This system, and the spawning grounds in relation to it, has been described by Lee (1952) and Corlett (1958). Lee (1959) has further analyzed variations in the volume transport of the West Spitsbergen Current and has shown the dependence of this current upon the development of the polar high pressure system. He has also suggested a possible mechanism of the cooling processes and anomalous advections in the area and has thus indicated that the change in the transport might be made predictable.

The seasonal depth migrations of a given species is different in different ocean regions in respect to both depth range and timing. These differences are illustrated in *Fig 33* with the seasonal vertical migrations of the Pacific cod in various parts of the North Pacific Ocean.

It has been observed that most of the seasonal depth migrations start relatively suddenly and their timing varies from year to year. It can be postulated that they are triggered by environmental changes, *eg* by relatively sudden, but small, change of bottom temperature caused by the mixing of the water column by the first autumn storms. This hypothesis needs verification by future research.

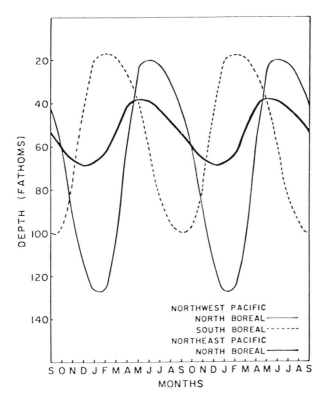

Fig 33 Schematic illustration of seasonal vertical migrations of Pacific cod (*Gadus macrocephalus*) in various parts of the North Pacific Ocean (from Ketchen, 1961).

The effects of seasonal depth migrations on the quantitative distribution of species biomasses can be of considerable magnitude. These quantitative effects can be computed with computerized migration models. One example of such computations is given below.

The seasonal depth migrations of flatfish were investigated by Alverson (1960) and on the basis of his work it was assumed that yellowfin sole migrate in the Bering Sea from deep water into shallow water during May and June, and back into deep water in October and November. A migration speed of 3km/day was assumed in the model. This resulting monthly change of biomass can be considerable in an area where active and extensive migrations occur (see *Fig 34*). These seasonal migrations have profound effects on other biota as well as on the evaluation of fishery resources with trawling surveys. For example, flatfish are dependent on benthos as a food source and migrations cause heavy grazing of benthos in some areas during some seasons, allowing a 'recovery period' during other seasons. A proper trawling survey evaluation must account for seasonal migration to avoid erroneous results.

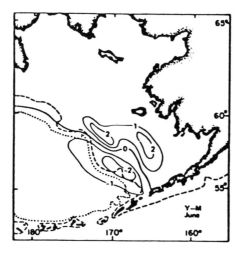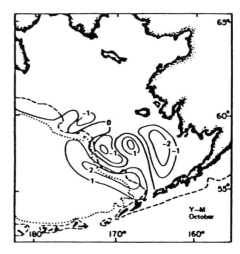

Fig 34 Change of yellowfin sole biomass distribution due to migration in June and October in the eastern Bering Sea (t/km²).

Seasonal migrations are affected by various environmental anomalies and, especially at high latitudes such as in the Bering Sea, water temperature anomalies are usually the most pronounced and the easiest to observe. However, it is not always possible to determine a single cause for seasonal migrations, which can be spawning migration, search for food, or search for acceptable or optimum environmental conditions.

6

Fishery in relation to environment

The proven and traditionally utilized relations between the fishery and the environment have been: the knowledge of seasonal fishing grounds, the depth of the water, and sometimes the nature of the bottom.

The effects of environment on the fishery of a given species are location and season specific. These relations have not been studied and/or summarized for any given species in sufficient detail to be fully useful for practical fishing. The reasons are that traditional fisheries research has been occupied with the study of biology, ecology, and resource evaluation for management and conservation. In this chapter we will present some available examples of the relations between fishing and environment.

Major factors, both environmental and biotic, affecting the exploitation of fishery resources, are systematized in *Fig 35*. Furthermore, the effects of various environmental conditions and processes on the fishery resources and their exploitation are systematized in *Fig 36*.

Determinant condition	State or process within resource	Affecting ecological and environmental process/condition
AVAILABILITY TO FISHERIES	Aggregation, dispersal, abundance in given area	Food availability State of maturity, seasonality Environmental anomalies (temperature, currents)
	Behaviour (diurnal, seasonal)	Environmental anomalies (thermal structure, light conditions, preferred food distribution)
FISHABILITY	Availability (abundance, concentrations)	See affecting factors above
	Fishability of grounds	Type and roughness of bottom Weather and sea conditions
	Behaviour in respect of gear	Diurnal behaviour Visibility of gear Interspecies interactions in respect to gear (e.g. pollock, salmon)

Fig 35   Major factors affecting exploitation of fishery resources.

Environmental process/condition	Effect on fishery resource
A. Effects on resources	
PAST THERMAL HISTORY (ANOMALIES)	Speed (time) of maturation (delay of spawning) Growth, including effects of food availability
THERMAL ANOMALIES	Displacement of spawning Growth of biomass (Availability of food for larvae)
THERMAL STRUCTURE WITHIN DEPTH	Availability, diurnal behavior (re. thermocline) Distribution and abundance of demersal fish
CURRENTS (WIND CURRENT ANOMALIES)	Transport (especially eggs, larvae) Mixing, transport of food (e.g. plankton) Effects of currents on distributions Advection of ice
B. Effects on exploitation	
THERMAL STRUCTURE	In depth distribution, diurnal behavior
STORMS	Fishability
TYPE OF BOTTOM AND DEPTH	Fishability
RESOURCE ABUNDANCE AND AVAILABILITY	(See A above)

Fig 36 Environmental effects on resources and exploitation.

6.1 Fishery in relation to oceanographic factors

Summary

Fisheries on many species occur in areas where fish aggregate due to environmental conditions such as temperature limits at thermal boundaries. Temperatures are often better co-ordinates for fish location than geographic co-ordinates. Furthermore, the thermal structure with depth determines fishing methods and tactics for many species.

Submarine illumination, which is greatly affected by turbidity of the water, affects the catchability of nearly all species, either via visual reaction to gear or via influencing the diurnal migrations and availability to gear. Night-light fishing of phototactic species is profitable only in relatively clear waters, such as in the Mediterranean Sea.

Currents (specially stronger tidal currents) affect the operation of all gear and catchability. However, due to short-term variability of currents and difficulties in measuring currents at sea, they are not fully utilized in fishing operations as yet.

The correlation between the catch of fish and the surface temperature has been the most studied environmental relation, probably because the surface temperature is the most easily measured oceanographic parameter. Magnusson *et al* (1979) have established and evaluated three independent postulates regarding the relation between fish and oceanographic features:

78

(1) temperature or other environmental axes are a better set of co-ordinates for finding fishes than is geographic space;

(2) man's view of fish habitat is often distorted because it is based on our own perceptual abilities and values, rather than those of the fish; and,

(3) an *a priori* approach to predicting where fish are located based on laboratory and field experiments on the causal relationships, should be more efficient than simply correlating with oceanographic features.

Lee (1952, 1956, and 1959) has shown that profitable catches of cod, in Bear Island waters, were not taken in water with a temperature below 2°C except in July to September, when the cod sometimes entered very cold water in order to feed on capelin and krill. In May to June and in October to December, good catches of cod were made on Bear Island Bank, when the temperature of the bottom water was 2-4°C, along the boundary between the warm Atlantic water and the cold arctic water; the cod were sometimes found in pockets of warm water surrounded by water below 2°C.

Rasmussen (1955) found that in Labrador waters the best catches of cod with long line in the surface waters were made when the temperature was 3-4°C, and that satisfactory catches could sometimes also be made when the temperature was 2·5-3°C. The best catches with the bottom long lines were made at temperatures of 2·1-2·5°C. If the bottom temperatures were below 1·5°C or above 4·0°C, the catches were considerably smaller. McKenzie (1934 and 1936) indicated that most of the cod on the Nova Scotia Bank are caught in water at temperatures of 0·5-7°C. Le Danois (1934) claimed that at temperatures of 3-5°C cod are very abundant, and at temperatures of 5-7°C haddock are plentiful. Further information on the behaviour of cod in relation to temperature is given by Hachey *et al* (1954) among others. Dietrich *et al* (1959) concluded that a correlation exists between the concentration of haddock and the temperature in the region of the Dogger Bank as soon as the total number of fish in this area and the differences in temperature reach certain levels.

Sullivan and Fisher (1953) found that in autumn and in the beginning of winter the fish select progressively lower temperatures, the change being slow in autumn but rather rapid in early winter. Then in spring, independently of changes in water temperature, the fish seek higher temperatures. The finding of Thompson (1943) that catches of cod, in relation to temperature, vary with the season near Newfoundland, is in accordance with this observation.

Thus, whatever the actual causes, the fluctuations in the catches of many pelagic species are 'controlled' by temperature (see Uda and Okamato, 1936; Nakai, 1959). The relative distribution of catchable stocks are correlated with temperature (Radovich, 1959). Several attempts have therefore been made to correlate the sea-water temperature with the annual catch of fish, for instance with that of 'iwashi' (sardine, anchovy, and round herring) in the Sea of Japan. In addition, Uda and Honda (1934) also concluded that the duration of the fishing season for 'buri' (yellowtail) (*Seriola quinqueradiata*) in the waters around Japan is longer when the temperature of March is lower than that of January, and also when the temperature rises more slowly than usual. Uda and Watanabe (1938) assumed that the passage of cyclonic storms causes the lowering of sea surface temperature and the southward extension of the Oyashio Current, resulting in a rapid southward shift of the fishing areas of skipper and bonito.

In order to be able to predict the presence of fish plentiful enough for profitable fishing in a given area:

(1) The optimum temperatures (and the optima of other environmental factors) of all economically significant species must be known.

(2) A sufficient number of frequent hydrographical and meteorological observations must be available to provide information on the location of critical surface isotherms and, furthermore, on the areas of sharp surface temperature gradients where pockets are formed by meandering eddies of the currents.

(3) The changes in the hydrographical conditions must be predicted.

Warning, however, should be given that the temperature where the highest catch is obtained may vary from year to year and from area to area. When observations from a large area and from a period of several years are handled together, the temperature range for the highest yield must be necessarily a wider one (eg 0·75 to 3·25°C for Greenland cod, Bratberg and Hylen, 1964).

The fishing operations as well as the distribution of fish can be affected by currents. (Some of the effects of the currents on fish behaviour were summarized in Chapter 2 §2.2.) In any fishing operation it is important to know the prevailing current, since the direction of trawling, the set of nets and other gear, such as the purse seine, are to a large extent influenced by the direction and speed of current. Often fishermen have their own tactics and tricks in those operations and it would not serve any purpose to describe them here, because one method and trick used in one condition might, and often does, bring undesirable results in other conditions.

Direct current measurements are difficult or even impossible to perform at sea, and their results are uncertain. Therefore, indirect conclusions must be drawn about the prevailing current from wind observations and tide data.

In trawling operations both the speed of the trawl through water and the area coverage influence the catches. Therefore, among other data, the actual shape of the trawl at different speeds and in different currents is useful information for the fishermen. However, information on the behaviour of the trawls at different speeds and in currents is scarce. The few studies performed during the last few decenniums concern mainly the resistance and shape of the trawl at different speeds and the influence of the length of the warps. It appears that the knowledge of the behaviour of different trawls in currents must be measured empirically because of the variations in construction and material which make the application of hydraulic formulae too complicated and uncertain.

Ketchen (1957) found that the opening of the bottom trawl was smaller when the boat was running with the tide than when it was running against the tide with the same engine power. Lower speeds increased the spread of the otter boards, but tended to decrease the height of the opening. Higher speeds tended to raise the trawl from bottom and to decrease the attack of the otter boards on the bottom.

The illumination in the sea, determined at any given depth by the diurnal changes of light and by the turbidity of water, affects the fishing with different gear. Gill nets and certain trap nets are in general more effective on dark nights and in turbid water, other conditions being equal. Fishing with hand and long lines for fish which locate their prey by sight, *eg* tuna, is most effective in clear water.

Murphy (1959) found that the number of albacore (*Thunnus alalunga*) taken by gill net on the surface was seriously affected by the transparency of the water. Observations assembled from a fishing cruise between Hawaii and the Aleutians in 1956, and from another cruise off the Pacific Coast of North America, showed that trolling was more effective in clear than in turbid water but gill nets were more efficient in turbid waters. He also suggested the possibility that water clarity may be an important ecological factor in the ocean in that the efficiency of sight feeders will be reduced as the turbidity increases.

Blaxter *et al* (1958) found that in complete darkness the herring passed through barriers which they avoided in daylight. This was taken as a proof that obstacles were avoided by herring with sight. On the other hand, herring in tanks were observed to swim occasionally along the barrier in darkness; it is possible that they were employing a tactile sense. Colours such as black and green, which presented the greatest contrast to the grey end walls of the tank, were found most effective as barriers. White was least effective. Transparent monofilament nylon acted as a very poor barrier. Obstacle avoidance by sight became ineffective at about 0·001 lux in light-adapted fish and at a lower figure in darkness-adapted fish.

Kanda *et al* (1958) found in their experiments that the variations in the behaviour of fish shoals near a coloured net were dependent more clearly on the wave-length of light than on the intensity of light reflected by the net twines. This observation can probably be explained on the basis of variation of the extinction coefficient as a function of the variation of wave-length.

Inoue *et al* (1958) have given a formula for the visual range for net twine in water of varying turbidity. Unfortunately, several factors in this formula depend on the observer's eye, the brightness of the background, and the visual angle of the tangent. The formula is therefore not applicable for many practical conditions.

Nomura (1958) found that when sardine shoals discover nets they sink into deeper water. Therefore, the Japanese fishermen set some nets deeper than the depth where they expect to find the sardine shoals.

Okonski and Konkol (1957) found that the herring shoals reacted to the mechanical disturbance of the environment if the disturbing tool attacked the shoal itself, like the pelagic trawl. Herring tried to escape, usually toward the bottom, and only occasionally toward the surface, or dispersed both upwards and downwards. Herring found in the sphere of the mouth of the trawl usually chose to escape upwards during the night;

however, under the intense light conditions of daytime the herring tended to escape downwards. How much of these reactions of fish to gear are directed by photic or tactile senses is still unknown.

The gill net fishermen nearly all over the world have recognized that the visibility of nets, combined with the turbidity of water and with currents are important in determining the fishability. Therefore, the nets are made of thinnest possible thread for a given mesh size and the preferred colour is usually bluish-grey. Furthermore, the catchability of gill nets during the night is related to the lunar calendar, the largest catches being usually made during cloudy nights or during the dark of the moon.

Schärfe (1953) gave a rather comprehensive review on the use of lights in the fishery up to the time of reporting. Of other works in this field the following could be mentioned:

Takayama (1949) described the Japanese saury lift net fishing with light. Dragesund (1958) gave an account of the light fishing in Norway for herring which started in about 1930. Today, both purse and shore seiners use lights to attract herring, especially the spring herring. The purse seiners are equipped with searchlights of 1 to 3kW which are switched on to attract the fish when the vessel is either at anchor or drifting.

Kawamoto (1958) pointed out that the individual gathering rate of different species of fish to the fishing lamp must be a function of the luminosity of the light source. The influence of the diurnal rhythm of the species, of the occurrence of fish food and predators around the lamp, of the intensity of moonlight, and of some hydrographical factors must be taken into account, when the gathering rate of a certain species is to be estimated.

Hsiao (1952) found that both yellowfin and little tunny were attracted to a continuously shining white light of *ca* 700 to 4 500 lux. They were repelled by stronger light. The fish were attracted to coloured lights with comparable intensities as well.

Uda (1959c) stated that the catch in the Japanese light fishing has increased in the last twenty five years almost in proportion to the second power of the light energy used.

Blaxter and Parrish (1958) found that in many cases it was apparent that fish found near the lights were predators attracted by the aggregation of prey.

Uda (1959c) sums up the response to light or phototaxis in respect to fishing lamps as follows: 'Phototaxis of ecological groups, as revealed by observations and sampling, shows the zooplankton responding first, followed by small fish, and finally by large fish, *ie* in the order corresponding to the food-chain (prey-predator) relationship. Fish shoals at the breeding or spawning seasons have, in general, no phototaxis.'

6.2 Fishery in relation to weather

Summary

Weather, specially surface winds, has profound effect on fishing operations (fishability). Winds affect also the availability of many fish. These effects are via waves and turbulence which affect vertical distribution and shoaling of fish. Most of the fish-weather relations are anecdotal and have been gained by fishermen during long periods of observations.

Only lately is some of this anecdotal information being checked with scientific approaches (such as wind and catch relations of Lowestoft trawlers).

———————————

The behaviour of fish (horizontal and daily vertical migrations, aggregation, dispersal, *etc*), and consequent accessibility of fish, are determined by a multitude of environmental factors which in turn are directly influenced and/or determined by the meteorological conditions. A good review of the known relations between the weather and fish stocks was made by Hempel (1960).

Fishermen know from their observations and practice that storms have great influence on the occurrence and migrations of fish. Some more scientific investigations into this subject are also available. Robins (1957) concluded that the storms are limiting factors in the shoreward distribution of many fish. He further assumed that species of fish, not normally found in shallow exposed waters, may become established in such habitats during periods of calm weather. The passage of storms or persistence of onshore winds result in turbulent conditions in shallow waters. Certain species may be unable to withstand the turbulent conditions and are killed, apparently due to the erosion of gill filaments by accumulated sediments.

Uda (1927, from Uda, 1959c) studied the catches of yellowfin tuna in the waters off Nagasaki Prefecture in Japan. He found that in winter and spring the catches were related to the passage of cyclones and atmospheric fronts. Maximum catches were made one to two days before and after the passage of a cyclone, possibly as a result of the turbulence in the sea. Similar statistical-meteorological correlations were found later in Segani Bay in Japan.

Heavy storms which usually cause lowering of surface temperature influence the catch (Lauzier, 1957; Uda and Watanabe, 1938). This lowering of temperature can be brought about in two ways:

(*1*)   by mixing of warm surface water with deeper cold waters,

(*2*)   by horizontal movements of surface waters bringing about intensified upwelling or sinking and thus causing vertical movements of waters of different temperature.

In shallow water storms cause high turbidity and limit the shoreward distribution of some fish which cannot withstand turbid conditions (Robins, *op cit*). However, certain trap nets catch best during stormy weather (*eg* salmon and eel trap nets in the Baltic Sea). Furthermore, even the gill net catches are higher in turbid waters after storms. However, the conditions on fishing grounds, and thus especially the catches of smaller pelagic fish, *eg* herring, can be drastically changed during and after storms.

It is also known that whaling is greatly influenced by weather conditions, probably indirectly. Heavy concentrations of whales often coincide with areas of dense advective fog, which is connected with convergence areas of currents with different temperatures and are areas of higher abundance of krill.

Corlett (1965) demonstrated that wind (and plankton) data can be used as a basis for the prediction of recruitment (year-class strengths) in some fisheries up to five years in advance (*ie* when a year-class becomes subject to fishery). Walden and Schubert (1965), having treated statistically a great amount of wind and catch data from the North Sea, found that wind direction and force are correlated with catch only in a few areas. The relations between the weather and the catch are usually complex and are often masked by other factors which must be taken into consideration in a forecasting service.

Harden Jones and Scholes (1980) investigated the relations between wind and the catch of a Lowestoft trawler. Their analysis of the data showed that over the year the catches of plaice were lowest with northerly winds but that the reverse was true for cod. For plaice, below average catches were associated with northerly winds in autumn and in winter; for cod, higher than average catches were associated with northerly winds in autumn and in winter, and with southerly winds in spring and in summer. The seasonal wind patterns suggested that the apparent relation between wind and catch might be coincidental rather than causal, winds from one direction being dominant during periods of high or low availability. Direct evidence of a causal relation between wind and catch was provided by a change of wind between trips being associated with a significant change in catch rate when the trawler returned to the same ground. The effect of wind on the catch of plaice was considered to be affected by swell and sea, oscillatory bottom currents, and turbulence.

Almost everyone's daily business depends on the weather, but there is scarcely any other profession which is dependent to such a high degree on it as that of a fisherman. The daily and hourly weather affects the behaviour of fish and their availability to fishing gear, resulting in relatively great variations in catch. The fishermen usually have to work in difficult conditions, being not only directly exposed to the weather, but simultaneously to many aspects of it. In the less developed fisheries, the losses of life caused by weather are relatively large.

The operations, even of large fishing vessels, are determined by weather conditions, so the catches and landings depend greatly on the frequency of storms during the main fishing season for a given species. Modern high-sea fishing vessels usually have to stop fishing at 7 to 8 Beaufort. The typical coastal fishing vessels in the North Sea find difficulty in fishing at 6 Beaufort, and coastal fishing by small boats must be stopped earlier. The fishing must sometimes stop at even a much lower wind force than indicated above, if there is an older swell present, which, by superimposing on the new sea caused by local winds, gives 'outsized waves' which endanger the fishing. Beach landing is often a limiting factor for a coastal fishery because of sea and swell conditions, although the conditions outside the surf zone might be reasonably safe.

The formation of ice on the superstructure of ships, called 'icing of ships', constitutes a menace to the safety of ships at sea and may thus seriously hamper the fishing in higher latitudes. In several cases this menace has become fatal and has caused the loss of fishing vessels and fishermen.

There are various kinds of ship's icing:
(1) Icing by sea-water—
    1.1 By water-splashes due to the interaction between the waves and the ship's movements.
    1.2 By spray being blown from the crest of waves.
(2) Icing by fresh-water—
    2.1 By freezing rain and/or drizzle, occasionally also by wet snow.
    2.2 By fog.

There are various means and methods of diminishing ice aboard a vessel; however, they are sometimes not sufficient to hinder its further accretion if the icing goes on rapidly or if working on deck is impossible owing to heavy seas and breakers. Ice accretion is a complex process that depends on ice conditions, sea conditions, and the ship's size and behaviour.

The weight of ice formed on a fishing vessel can be large and can reduce the freeboard. As ice formation and accretion take place on the upper parts of the hull, at the superstructure and at the rigging, thus elevating the center of gravity and making it topheavy with diminished stability, capsizing may follow. Moreover, the antennas may break down causing an interruption of radio communications. Instruments important for navigation may cease operating due to over-icing. Owing to the enlarged surface of the ship's upper parts by accumulating ice, the vessel drifts with the wind at a higher rate; therefore, handling the ship may become difficult.

The air temperature below the freezing point of water is the basic condition for the formation of ice on board. Cold air cools the ship's hull and its superstructure below the freezing point of fresh-water ($0°C$) or of sea-water (normally for colder parts of the ocean $-1·8°C$). Thus water particles striking parts of the ship and adhering to them begin to freeze. Freezing spray is the most common and dangerous form of icing of ships.

The following meteorological situation has been called by the British Shipbuilding Research Association in their trawler-icing research 'typical' for the icing of ships by sea-water:

Air temperature	-4 to $-7°C$
Wind force	6 Beaufort or more
Temperature of sea-water	$+2$ to $-2°C$

The menace from icing is at its maximum in stormy conditions coinciding with severe frost. On the other hand, air temperatures below $-18°C$ cause the sea-water droplets to freeze in the air before reaching the ship. Such conditions are far less dangerous.

In addition to the above various weather elements affecting the fishing, from which the especially dangerous and unpredictable icing was referred to at some length, it must be remembered that almost all kinds of fishing gear are adversely affected by waves and currents in the sea. It is worth noting that the weather places similar restriction on the use of many kinds of oceanographic instruments and sampling gear for research purposes. Naval architects now take seriously into consideration the prevailing sea and weather conditions on the fishing grounds when designing appropriate fishing vessels for operation in given areas. For this reason the investigations of sea behaviour and sea-worthiness of fishing vessels has been intensified in the last few decenniums.

The handling and preservation of fish is directly affected by weather, especially at sea. For various reasons there is a tendency to operate offshore fishing vessels during different seasons on different fishing grounds, with great latitudinal range. Important fish handling and storing problems on board are a consequence of the different climates on these different grounds lying far apart.

The present weather services for shipping and fishing, provided by many maritime countries in higher latitudes, more or less cover the present-day needs for fishing operations. There is, however, still much leeway for improvement in many areas, and the future will pose new requirements.

Specially prepared meteorological information can be of great use to fisheries scientists for making prognoses, *eg* forecasts of the strength of the year-classes and of the fishing conditions in various areas. The success of spawning and the number of young fish surviving greatly on the prevailing meteorological conditions during and after the spawning, and from properly prepared meteorological data (*eg* cumulative wind data) the drift of eggs and conditions for development and survival of fry can in some cases be estimated. These estimates need, of course, to be checked by sampling surveys with research vessels, but considerable savings on the surveys can be made and the accuracy of the prognoses increased by using the results of calculations based on meteorological parameters.

7

Fish finding in relation to environmental factors

7.1 Fish finding using the distribution of environmental factors

Summary

Fish scouting with 'thermometric methods', *ie* utilizing the knowledge of temperature requirements of given species together with quasi-synoptic analyses of temperature, can cut searching times in some fisheries. Scouting for demersal fish requires the knowledge of bottom temperatures. Knowledge of thermal structure with depth, obtainable with an XBT, is used to determine the depth of fish (and shoals) when the diurnal behaviour of the species is known.

Current and water type boundaries, which are often also thermal boundaries, are often areas of aggregation of fish. These boundaries are detected from the changes of colour, appearance of sea surface ('modified waves'), and from sea surface temperature gradients.

Several environmental factors may interact to determine the distribution and availability of fish. In the past the relation between fish availability and a single factor has been emphasized, but it appears desirable to investigate the compound influences in the future.

If there are meaningful relations between the occurrence of fishable concentrations of fish and some easily observed environmental parameters, it would be possible to delineate the areas of search for fish from environmental analyses/forecasts, thus cutting wasteful scouting time. For many commercial fish species, such meaningful relations have been found. For example, optimum temperature range for the cod in the northwest Atlantic is 2° to 3°C during the winter and 3° to 5·5°C during the summer. Thus, if synoptic sea temperature analysis charts are available, the general areas for possible occurrence of cod can be delineated, especially during the winter, when the sea is isothermal from surface to bottom over many continental shelves and shallow seas. A hypothetical example of this is shown in *Fig 37*. After the general area of profitable fishing has been outlined through use of temperature criteria, further definition can be achieved by applying other knowledge such as depth of the water, type of bottom

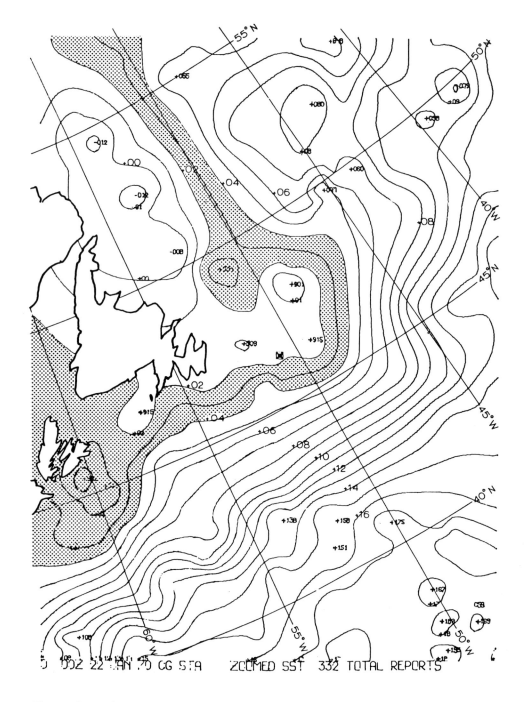

Fig 37   Sea surface temperature analyses off Newfoundland-Grand Banks area on 22 January 1970 (the areas with temperatures 2 to 4°C are hatched).

preferred by fish, trawlability of the bottom, and the historically known (or traditional) fishing grounds.

While outlining of potential fishing areas from temperature charts may not present any new and useful information over and above the knowledge of traditional seasonal fishing grounds, knowledge of other influencing factors such as temperature anomalies and their persistency might provide additional useful information. One such example is schematically shown in *Fig 38*.

Haddock spawn in April at the continental slope within a temperature range of 5° to 7°C (upper figure). In a given year, temperature conditions on the normal spawning ground may be too cold for spawning (lower part of the figure). Thus, good fishable concentrations of haddock would not be expected there. The spawning areas might have been displaced horizontally or the fish might spawn in deeper layers and not be accessible to fishing. Furthermore, there may be a delay of spawning activity, especially if there was a large-scale negative temperature anomaly in the general area (which we can

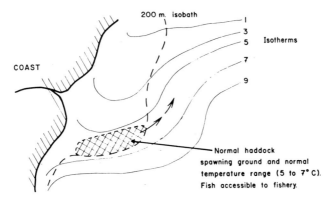

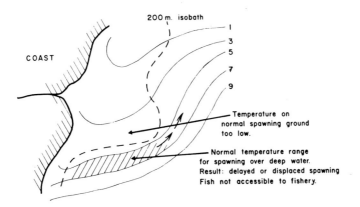

Fig 38   Schematic example of the effect of negative temperature anomaly on haddock spawning and fishing.

89

ascertain from past synoptic analyses and anomaly computations); thus, a late season for the fishery may be expected.

Although the temperature is the environmental factor most easily measured, and likewise the most important one influencing the behaviour of fish, many other factors may be involved; thus, predictions of fish concentrations based on temperature cannot always be expected to be correct. Such predictions are more likely to produce better catches if the following conditions are fulfilled:

(1) The behaviour of a given species at various temperatures must be known (including optimum temperature and spawning temperature). Much of the available information on this subject is brought together in Chapters 2 and 3, from which it can be extracted for the particular species of interest.

(2) The present distribution of temperature must be known and the future course of its distribution must be predicted.

(3) In addition, it is necessary to assume that the horizontal (and/or vertical) temperature gradients are not too weak.

The following examples will illustrate the present use of synoptic oceanography to fisheries.

Shapiro (1950) stated that in Japanese long line fishing for tuna the data on the locality for the best catch, when correlated with information on optimum water temperature for tuna, configuration of ocean bottom, and type of ocean current, have proved useful in indicating to fishing vessels the situation under which the long line gear can be operated to maximum advantage. Knowledge of currents, convergences, divergences, reefs, banks, and islands is also important to Japanese fishermen when ascertaining shoals of tuna and other fish. The Norwegian biologists (Devold, 1951 and 1959), in intimate co-operation with their hydrographer colleagues, have made use of the close correlation between the temperature distribution and the migrations of the herring for the prediction of the latter. The winter herring, approaching the Norwegian coast, often gather between Iceland and Norway in cold water pockets before penetrating the warm Atlantic Current into the colder Norwegian coastal waters. By following the surface temperature in this area during the second half of December and the first half of January, the offshore fishing fleet can be directed to the fish concentrations in these pockets. Also the arrival of herring in the coastal area can be predicted with greater accuracy which means considerable economic gains.

The sea surface temperatures during the above two periods (of the winter 1958-59) are shown in *Fig 39*. The quasi-synoptic temperature charts have been constructed by Norwegian fisheries scientists utilizing not only the data from fisheries research vessels but also the meteorological reports from observing merchant marine vessels. Thus the actual movement of the two cold water pockets can be followed in time and space. The one nearest to the Faeroe Islands was more pronounced, and the herring aggregated mostly in this pocket. The herring were first observed on 9th January by the research vessel *G O Sars* and followed until reaching the Norwegian coast in the Stad area on

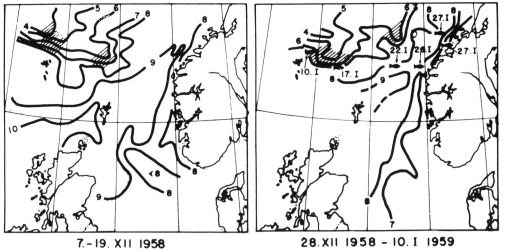

7.-19. XII 1958   28.XII 1958 - 10. I 1959

Fig 39 Occurrence and movement of herring schools in relation to surface temperature in the Norwegian Sea (after Eggvin, 1963). Concentrations are shaded; dates show locations of migrating shoals.

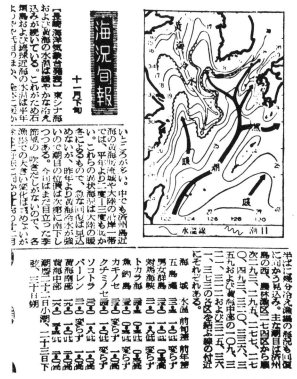

Fig 40 Example of a weekly surface temperature and currents chart for information to fishermen, from a Japanese newspaper. Warm and cold currents are indicated with different arrows. The convergence between cold and warm water is shown with a wiggly line.

25th January. The speed of migration seemed greatest where the Atlantic Current was strongest.

The use of synoptic oceanography for fisheries is most developed in Japan. A weekly combined surface temperature and current chart, published in newspapers, is given as *Fig 40*. The convergence line shown on the chart is a most important fishing area during the winter season. The knowledge of the location and movements of this convergence is of utmost significance for the fisheries. Synoptic data for construction of the charts and forecasts are obtained from commercial fishing vessels and from research vessels.

Jakobsson (1971) has described the complexities in the effects of temperature changes on the herring migration in the Norwegian sea, demonstrating that finding of the herring with the surface temperature alone might not always give desired results. The herring often migrated in small scattered shoals across the East Icelandic Current in late May and early June, especially if a relatively warm surface layer had developed above a sharp thermocline. In the years between 1965 and 1968 the increased portion of polar water in the East Icelandic Current brought the sea temperature in this area below a tolerated threshold level and the herring in May to June 1965 entered the cold water and soon turned north and northeast in search of more viable environment. In other years when the sea temperature is above such a threshold level, the temperature has less apparent effect on the herring distribution. In such cases the feeding conditions in the late spring and early summer seem to have a greater effect. Thus the distribution of the herring may in some years be mainly affected by one environmental factor but in other years the apparent effect of a different environmental factor may be of more importance.

The vertical temperature gradients in the sea are several orders of magnitudes sharper than horizontal (*eg* sea surface) temperature gradients (see further Chapter 8). The thermal structure with depth is usually measured with bathythermograph (BT and/or XBT), but also may be measured with a temperature sensor in a netsonde attached to a trawl. The use of the information, obtainable from a BT, in fisheries problems varies considerably from one type of fishery to another. Therefore, only a general list of some fisheries applications of the knowledge of thermal structure with depth is given below:

(1) There are pelagic fish which are found above the thermocline, and still others which are found mainly in deep water (see *Fig 41*). Thus, the information on the depth of the thermocline can be used for:

 (*a*) setting the depth of the long lines (*eg* for different tuna species),

 (*b*) setting the depth of the drift nets (*eg* in herring and salmon fisheries),

 (*c*) determination of the optimum depth of midwater trawling (see *Fig 42*),

 (*d*) deciding whether a purse seine cast is advisable [*eg* in case of a deep mixed layer depth (MLD) shoals might dive below the pursing depth (see *Fig 43*)].

(2) Many species have diurnal vertical migrations, which are limited either upwards or downwards by the existence of a sharp thermocline which forms an environmental barrier. Thus, the knowledge of the thermal structure and of the depth

of the surface mixed layer (MLD) is useful when determining the fishing tactics for such species.

(3) There are species which aggregate in the thermocline regions and especially in areas where the thermocline would intersect the bottom off a coast. A knowledge of the thermocline depth provides a means for the discovery of these fish.

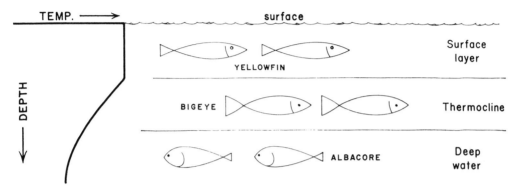

Fig 41 Schematic example of different depth and temperature preference by different species of tuna in tropical latitudes.

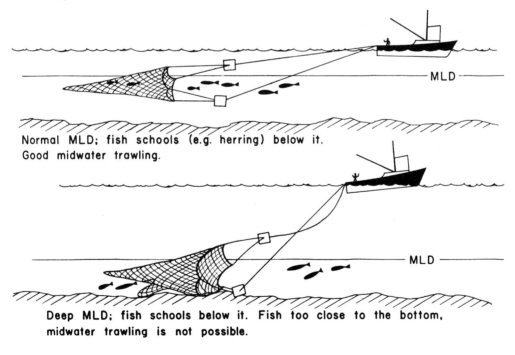

Normal MLD; fish schools (e.g. herring) below it. Good midwater trawling.

Deep MLD; fish schools below it. Fish too close to the bottom, midwater trawling is not possible.

Fig 42 Schematic example of the possible effect of the mixed layer depth on midwater trawling.

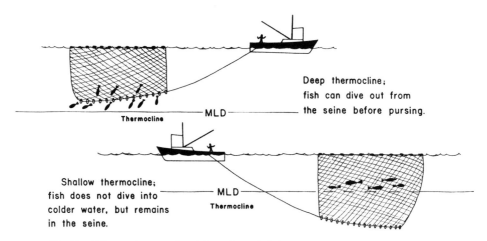

Fig 43 Schematic example of the effect of different mixed layer depths on purse seining for pelagic fish.

(4) Most species prefer certain optimum temperatures. Moreover, their normal distribution is limited between a minimum and a maximum temperature. Since the sea surface temperature (SST) does not provide information about the temperature at a preferred depth for a species, a BT cast is necessary for the determination of a possible occurrence or lack of fish in a given area and depth, especially near the limits of their geographical distribution (*eg* in the Bear Island and Spitsbergen area). (See *Fig 44*).

(5) Under certain conditions there is an accumulation of species near the current or water type boundaries. A water type boundary might be considered as a limit of the normal distribution of a given species. The BT casts provide the best information on water types and their boundaries and, therefore, on the location of fish shoals.

The most significant horizontal temperature gradients in the sea occur in regions of divergence and convergence of currents. These current boundary areas can be rich in pelagic fish. Therefore, the prediction of the positions and movements of these zones, with the meandering and eddying of water masses at their boundaries, is of great importance for the fisheries. This is also true when searching for whaling areas (Hanzawa *et al*, 1951). The relations of the current boundaries to the fish concentrations were further discussed in Chapter 3 §3.3. Two additional references to this subject are given below:

Beverton and Lee (1965) showed that on the Spitsbergen Shelf the cod was confined to the Atlantic water warmer than 1·5 to 2·0°C and that their distribution varied in accordance with the position of the boundary between warm and cold waters, both on a broad scale and even locally, *eg* when a sharply defined temperature structure built up during the warm spring of 1954.

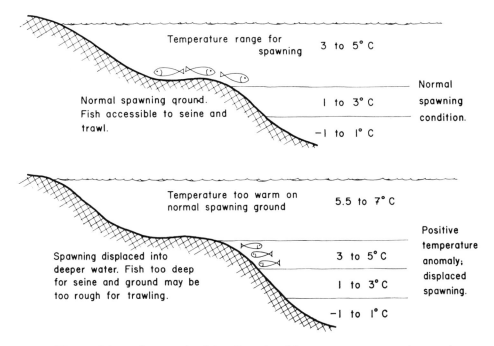

Fig 44 Schematic example of the effect of positive temperature anomaly on cod spawning and fishing.

Jakobsson (1971) stated that in most cases the densest herring concentrations in Icelandic waters during the years 1957 to 1966 were located at, or just outside, the continental shelf in the boundary areas of warm and cold water masses. Furthermore, the concentrations usually extended along the current boundaries but had a less extensive distribution pattern in other directions.

The current boundaries keep changing positions with the seasons and with varying meteorological conditions as well. Predictions of their movements and future positions is an important subject of synoptic oceanography. The boundaries are usually characterized by the change of a few or of all the following properties, depending upon the nature of the boundary:

(1) sharp surface temperature gradients,
(2) water colour change,
(3) bands of calm and rough water,
(4) type of sea, resulting in a confused sea,
(5) whirlpools,
(6) irregular steering of ships (which can go out of course as much as 10° if steered by the gyro pilot).

Sometimes the determination of salinity – with the temperature – and the examination of plankton samples are necessary for determination of water-masses and their boundaries.

95

In addition to the current boundaries and the temperature, a variety of other hydrographical indices for the determination of favourable fishing grounds of a given species can be listed. These indices vary from species to species and from area to area. They must be established by local fisheries research workers. As an example for such criterions for favourable fishing grounds – and periods – those for squid in Japanese waters are given below as listed by Uda (1959c):

(1) Sandy bottom, mixed with shells and gravel, or flat rocky reefs or banks, 30 to 100m depth.
(2) Near the slope of the continental shelf.
(3) Squid sink to the bottom during the day but are in surface layers during the night.
(4) Good catches are made during cloudy, calm, and warm declining weather. Rainy and windy weather and strong tidal currents make the squid sink to deeper layers.
(5) If the vertical temperature gradient is around 5°C per 100m, the catch is good; if the gradient is too big or too small, the catch is poor.
(6) Good catches are made in eddies with upwelling.
(7) There is a tidal period in fishing, with the slack of tidal currents corresponding to the best catches.
(8) The time of full moon is in general a time of poor fishing, since the effectiveness of artificial lights used in fishing is reduced.

Uda (1936) stated that saury (*Cololabis saira*) stay in the area between the boundaries of the Kuroshio and Oyashio Currents and that they always migrate in the direction of the maximum gradient of the surface temperature. According to Uda (1952) the best areas for pelagic fishing are in general the convergences of currents.

The relation between a convergence and fish aggregation might not be a simple statistical one. Three hypothetical conditions might be recognised in this relation:

(1) The converging currents as such might be weak, below the threshold speed for the orientation of fish. But zooplankton is slowly accumulated at the convergence and the aggregation of fish might occur as a result of the ample food present at the convergence. This way the convergence might act as an environmental boundary indirectly.
(2) The converging current might be of medium strength; the fish orient themselves to the current and aggregate at the 'up-current' end of the convergence. Slight accumulation of zooplankton occurs as well.
(3) The converging current might be strong; the fish head into the current but are carried along with it. Fish aggregations are possibly observed at the 'down-current' end of the convergence. Because of the strong current, no aggregation of zooplankton is possible.

Several hypotheses on the effects of currents need testing with observational data.

Many species are caught in greater quantities during their aggregation for spawning.

The arrival of spawning stocks as well as fluctuations in area and time of spawning can be predicted relatively accurately in many species, if both the past and present history of the temperature in the sea and the specific thermal requirements of the species are known.

In order to be able to map the sea temperature distribution in detail in a short period (such as five days or a week), systematic hydrographic observations, supplemented by some meteorological observations, are necessary. In the most important fishing areas of the world this goal will be attainable if and when suitable steps are taken to organize the collection of the information needed (further see Chapter 10).

During the past sixty to seventy years oceanographic and fisheries investigations have provided a variety of knowledge on the behaviour of fish in respect to the environmental conditions met in the sea. However, little use has been made of synoptic environmental analyses in fisheries research in the past. Two reasons for this are that (a) fisheries research has been biologically oriented, rather than application oriented, and (b) synoptic environmental analyses/forecasts have not been available to fisheries.

One of the main shortcomings in the use of environmental data in fisheries research has been the attempt to analyse the effects of only one environmental feature on fish behaviour and distribution. The interaction of the fish and the environment is extremely complex and it is obvious that a number of features and processes are affecting fish populations. The availability of synoptic data and forecasts, and the use of high speed computers permits analyses of a number of interacting factors and processes resulting in a more realistic appraisal of fish behaviour and distribution. Furthermore, by asking a number of related questions pertaining to possible interaction, more applied information is obtained. There is a need to work up the various accumulated fish-environment relationships in computer compatible form. After this has been done, application of synoptic oceanographic analyses and forecasts may provide useful and timely information on fish behaviour and distribution. Finally, it could be mentioned that the availability of environmental analyses/forecasts will enable a rational planning of fisheries research, *eg* location of sampling and monitoring stations and determination of frequency of sampling and/or observations.

7.2 Effects of temperature and other environmental factors on fish finding with sonar

Summary

Temperature, salinity, and pressure determine the sound speed in the water, and the temperature has the greatest effect. Thus thermal structure with depth determines the presence of surface sound channel, its thickness, and the subthermocline 'shadow zone'. The transient thermoclines influence the efficiency of fish detection with hull-mounted sonars.

The active fisheries sonars operate in noise-limited mode. The propagation loss is determined mainly as spherical spreading and is affected by surface wave conditions. In shallow water the bottom scattering loss has a great influence on sonar performance.

The identification of sonar target (echoes) is subjective and very uncertain; therefore, sampling of the target is necessary. It seems to be necessary to develop a small sampling gear which can be operated at high speed in desired depth (*ie* targeted to sonar echoes). Furthermore, fisheries sonar predictions (range and propagation loss) seem to be desirable in some areas.

The offshore fishing in the North Atlantic is unprofitable without sonar assistance. Several aspects of submarine acoustics have become a direct concern of fisheries where echo sounders and sonars are used. This chapter will mainly review the submarine acoustics in respect to the environment, the main subject being the sound transmission. Also the problems of scattering and absorption of sound energy, fluctuations of signal and environmental noises will be included. We shall be concerned mainly with active sonars and with frequencies between 5 and 80 kilohertz.

Submarine acoustical devices, echo sounders and sonars, are used in fisheries for the sounding of depths, for the detection of fish, to determine the position and behaviour of gear and fish, and for the survey of fishery resources.

Many of the developments and studies of submarine acoustics for various naval problems are applicable to fisheries problems and *vice versa*. A number of excellent books and summaries on submarine acoustics have been published in recent years, to which reference is made for further details [(Guieysse and Sabathe (1964), Simonsen Radio (Simrad) (1964), Tucker (1966), Urick (1975), Tolstoy and Clay (1967), Forbes and Nakken (1972), Cushing (1973), and Clay and Medwin (1977)].

Sound speed in sea water depends on temperature, salinity, and depth. The influence of temperature, salinity, and depth on sound speed can be illustrated with the following approximate numbers:

(1) 1°C increase of temperature (T) corresponds approximately to 4·5m/sec increase of sound speed.

(2) 1‰ increase of salinity (S) corresponds approximately to 1·3m/sec increase of sound speed, and

(3) 100m increase of depth (D) corresponds approximately to 1·8m/sec increase of sound speed.

If we assume the salinity to be 35‰, the approximate sound speed values can be taken from the nomograph on *Fig 45*. This nomograph can be used, for example, for conversion of BT (bathythermograph) traces to sound speed profiles.

An example of sound speed profile at a location in the Norwegian Sea is shown in *Fig 46*. *Figure 47* shows two typical sound speed profiles from surface to some 400m, where essential features of the sound propagation are shown. The surface sound channel (or duct) reaches from the surface to the top of the thermocline, where the sound speed is

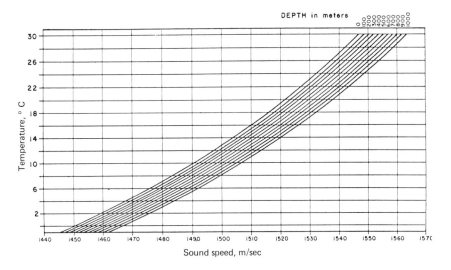

Fig 45 Nomograph for determination of sound speed from temperature (°C) and depth (meters) at salinity 35‰.

usually at its maximum. This sound speed maximum and the surface sound channel may be absent if the temperature decreases from the surface to the top of the thermocline.

In some limited ocean areas there exists a subthermocline duct, caused either by cooler, less saline water (North Pacific) or by nearly homothermal water (Sargasso Sea). A deep sound channel (main channel) exists in all ocean areas. The axis of this channel might be at the surface in very high latitudes during the winter. It can be as deep as 1 500m in low latitudes.

Sound waves, just like the light waves, are refracted from a denser medium to a less dense one, *ie* from higher sound velocity to lower sound velocity. The most commonly used theory for explaining the propagation of sound is the ray theory. The energy flux propagates along rays (ray tubes), the density of the flux (intensity) being proportional to the distance between rays. Using this theory, the modes of sound propagation in the sea are shown as *Fig 48*.

With the corresponding sonar geometry in shallow waters, it can be noted that the convergence zone propagation mode has no direct application to fisheries because of short sonar ranges. As the acoustic rays bend toward lower sound velocity, partial shadow zones are formed in certain distances below the depth of maximum sound speed in case the source is above this depth, and *vice versa* if it is underneath (*Fig 49*). These partial shadow zones are not entirely void of sound, as they receive some energy due to the reflections from rough surfaces and bottoms. The intensity of sound in the sea fluctuates considerably. These fluctuations are to a large extent caused by nonspecular reflections from wavy sea surface and rough bottom, by internal waves in sharp sound speed gradients (at the thermocline) and by random inhomogeneities in the water.

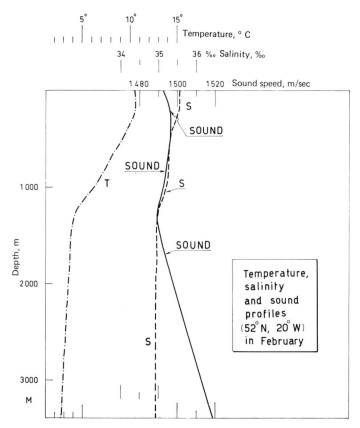

Fig 46 An example of temperature, salinity, and sound speed profiles in the northeastern Atlantic (52°N, 20°W) during February.

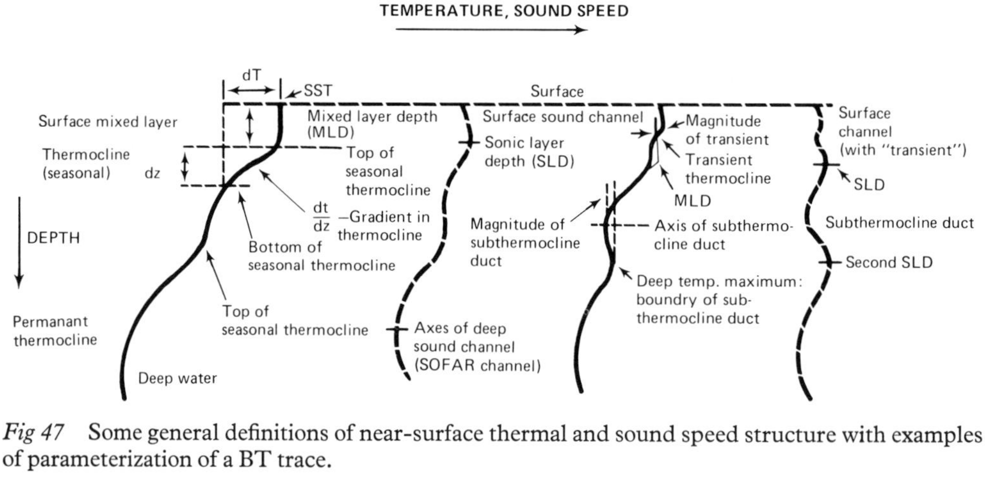

TEMPERATURE, SOUND SPEED

Fig 47 Some general definitions of near-surface thermal and sound speed structure with examples of parameterization of a BT trace.

100

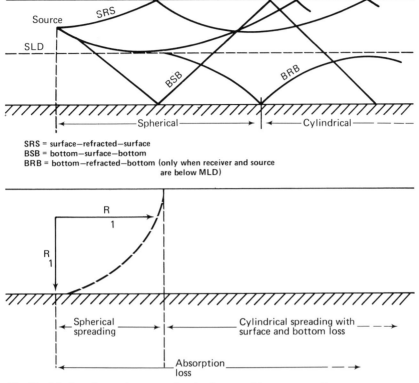

Fig 48 Modes of sound propagation in the sea with corresponding sonar geometry in shallow water. R is range and SRS, BSB and BRB indicate propagation modes. SLD – sonic layer depth; MLD – mixed layer depth (in most cases identical to SLD).

In order to explain the details of the effect of environment on the sound propagation and the use of sonar gear, the basic sonar equations must be understood. There are two slightly different active sonar equations, depending on whether the sonar condition is noise limited or reverberation limited. The noise limited situation occurs when the noise background due to the initial transmission of the pulse has disappeared and the noise (N) (ambient noise) present is that outside of the transducer (ship noise, environmental noise, *etc*).

The noise limited active sonar equation is:

$$M_n = (S_0 + T - 2H) - (N - \Delta R) \tag{4}$$

M_n is the signal to noise ratio, also called detection threshold. A target can be recognized on an average of 50% of the time if M_n exceeds a given number. S_0 is the source level and T is the target strength (or scattering cross-section) of the target. H is the one-way propagation loss. ΔR is the directivity index, *ie* the gain of the signal due to receiving directivity.

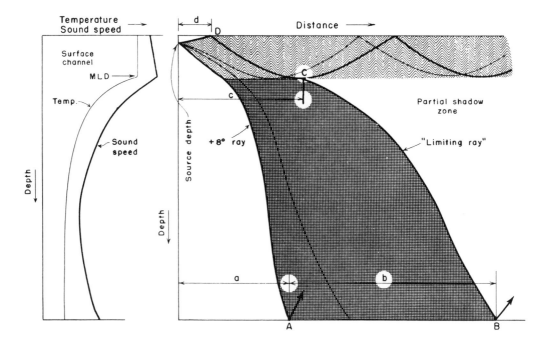

Fig 49 A typical temperature and sound speed profile and significant sound rays for a horizontal sonar beam. a – range of shadow zone on the bottom below ship, b – range of sonified bottom, c – range to shadow zone below MLD.

The reverberation limited conditions exists when the echo returns before the reverberation background has decayed to a level of the ambient noise.

The reverberation limited active sonar equation is:

$$M_r = (S_0 + T - 2H) - Q \tag{5}$$

where Q is the reverberation level present.

The source level is the intensity level at a reference distance of 1 metre from the source transmitted by a source. The target strength is defined as the ratio of incident to reflected intensity, referred to at 1 metre from the source. The noise level, N, depends on environmental noises, as caused by breaking waves, by noise-making marine animals, rain, traffic, and first of all by the ship itself carrying the sonar. The peaks in the noise spectrum of a ship vary considerably from ship to ship and according to the speed of the ship and the activity on board (*eg* the use of bilge pumps). The best way to increase the range and sensitivity of a sonar is to take measures to decrease the ship's own noise level.

The sound impulse emitted by a sonar or an echo sounder is subject to a number of losses during its travel through the environment: the absorption loss in water, the geometrical spreading loss, the surface scattering, volume scattering, and the bottom reflection or scattering loss. The propagation losses occur both ways, from the sonar to the target and back to the sonar receiver. Only part of the sound is reflected by the target and returned as an echo.

In a general case, we can assume a so-called spherical spreading of sound. In this case the propagation loss (in decibels, dB) is given with formula:

$$H = 20 \log R + aR + 10 \log \frac{\alpha}{\alpha_1} + SR + BR + VR \qquad (6)$$

In this formula the first term, $20 \log R$, is the true spherical spreading loss where R is the range; the second term, aR, is the absorption loss where a is the absorption coefficient; the third term, $10 \log \alpha/\alpha_1$ is the geometrical spreading loss; the fourth term, SR, is the surface scattering loss; BR is bottom scattering loss; and the last term, VR, is the volume scattering loss which is due to the scattering of sound from small marine organisms such as zooplankton, and caused partly also by sharp discontinuities in the environment.

The propagation of sound could best be illustrated in terms of rays, as is often done with the light where rays are the tangents to the wave fronts. In order to understand the bending of rays, one must know the distribution of sound speed with depth. The sound speed in water is a function of temperature, salinity, and pressure, the temperature playing the most important role.

The sound waves bend towards the lower sound velocity. Thus, with the sound source in the surface duct, the rays bend towards the surface, reflect back from it, and bend again at the so-called sonic layer depth (SLD) which usually coincides with the top of the thermocline. If the ray angle is above a certain value, the rays do not bend back to the surface but continue towards greater depths. However, due to the increase of the sound speed with pressure, some of them will again bend at greater depths and will return to the surface layers in so-called convergence zones. So far, the convergence zone propagation modes have little application to fisheries because of the large absorption losses involved in relatively high frequency sources. The only exception to this rule is the so-called half channel propagation (which is essentially a convergence zone propagation), caused during the winter time in high latitudes as well as in the Mediterranean Sea.

The different rays bend to a differing extent. Thus their spreading is no more radial at a given distance from the source. They can converge or diverge. The most important zone for diversion is at the top of the thermocline and near to it. This divergence causes the so-called divergence loss of propagation which can be computed very laboriously by hand methods but with relatively great ease on computers.

A systematic general presentation of ray trace diagrams, corresponding to specific bathythermograph profile (BTP) types which are useful for advanced planning by fisheries, is not normally readily available. Two selected ray trace diagrams for direct path sound fields are presented here (*Figs 50* and *51*). The source and receiver depth is 6m (20ft), corresponding to a hull-mounted sonar. The results shown cover only the upper 150m (500ft).

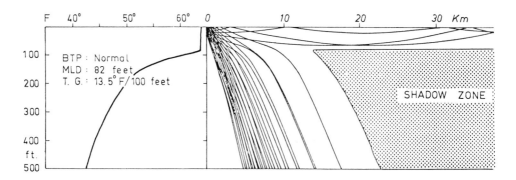

Fig 50 Ray trace diagram with indication of partial shadow zone for the normal thermal structure type: mixed layer depth 25 metres; thermocline gradient: 7·5°C/30 metres (13·5°F/100ft).

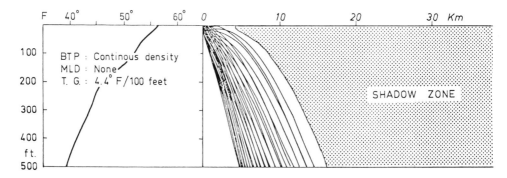

Fig 51 Ray trace diagram with indication of partial shadow zone for the continuous density type; no mixed layer depth; thermocline gradient: 2·4°C/30 metres (4·4°F/100ft).

The ray trace diagrams shown present only a general picture of sound propagation and sonified field for two BTP types. (In these two figures the depths are given in hundreds of feet and the horizontal distances in kilometres (or kiloyards). In the corresponding BT traces the temperatures are given in Fahrenheit.) In general, they do not depict, directly, the propagation loss or the sonar range. Those values depend on the frequency used, on the type of sonar, on the roughness of the sea surface (*eg* wave height), and on a number of other related factors, which for their part depend upon the particular sonar used. It should be noted that the sound intensity is approximately proportional to the density of rays (*ie* the distance between them) in a given distance from the source.

The zones of sound shadow, indicated by dotted areas, are not necessarily entirely void of sound. There is a diffraction from the advancing rays of the sound in the lower part of the surface duct and especially due to the reflection from a rough surface as the actual sea surface cannot be considered a mirror. Therefore, the attenuation in the surface layer is largely a function of wave height or of another roughness parameter, such as wind speed.

A comparative study of the ray trace diagrams in *Figs 50* and *51* demonstrates among other things:

(*1*)   the relative importance of the MLD,

(*2*)   the relative importance of the gradient in and below the thermocline,

(*3*)   the relative importance of the gradient near the surface, and

(*4*)   the difference in sound behaviour corresponding to different types of BT traces.

It is apparent that the MLD exerts the greatest influence on the extent of the sonified field near the surface. The gradient below the thermocline mainly influences the distance to the partial shadow zone below the MLD.

The influence of transient thermoclines on hull-mounted sonars depends directly upon their intensity and depth. In general, they may improve the range below the thermocline, but they can also have the same unfavourable effect on the range as a shallow thermocline within the depth of the transient. In some cases below the thermocline a thin sound channel is formed and can be utilized with the receiver and source in this channel.

If the velocity gradient with depth is negative (*Fig 51*, continuous density type), no true surface sound channel is present, and the sonar range is greatly dependent on the vertical temperature gradient.

An absolutely smooth horizontal sea surface would reflect the sound waves like a mirror reflects light. However, the sea surface is very seldom smooth and horizontal because of the waves. Thus sound which is bouncing to the surface is scattered in various directions due to the inequalities of the surface. This scattering is obviously dependent on the surface wave heights as well as on the frequencies of the sound waves. The only reasonable approach at present, for accounting for the surface scattering loss is by the use of empirically derived relations between the loss, sonar frequency, and wave height:

$$S = 1 \cdot 5 \sqrt{Wf} \qquad\qquad (7)$$

The scattering loss (S) in dB per kiloyard (or kilometre) is given in a one-way propagation. In this formula, f is the frequency and W is the wave height in feet.

The scattering of sound by the bottom depends on the frequency of the sound, on the sound speed in the sediment, which is a function of the porosity of the sediment, on the angle of incidence, and on the roughness of the bottom. In practical applications at sea it is difficult, if not impossible, to account for all of these parameters with sufficient accuracy.

In theoretical consideration there is a critical angle of reflection which is greatly dependent on the relations of sound speed in the water close to the bottom and that in the sediment. This critical angle varies from 30° in the case of mud to 70° in the case of rock.

Most of the fishing of the world is exercised over the continental shelf and slope, which for sonar purposes may be considered shallow water. The sonar conditions in shallow water are considerably more complex than over deep water. At the present time we can account for the shallow water propagation loss only in a relatively empirical way.

The different modes of propagation in shallow water were shown in *Fig 48*. For short ranges, the following propagation loss formula applies:

$$H = 60 + 20 \log R \tag{8}$$

and for long ranges, the following formula is applicable:

$$H = 60 + 20 \log R_{spherical} + 10 \log (R/R_{spherical}) \tag{9}$$

where H is the propagation loss (in dB), R is the range (distance) in km, and $R_{spherical}$ is the range for spherical propagation.

The following guides might serve a purpose for rapid estimation of propagation loss:

(*1*) If the range (R) is up to 1·5 times the depth, use 20 log R propagation – the spherical propagation.

(*2*) If the range is twice the depth or more, use cylindrical – the 10 log R propagation as in formula 9.

(*3*) If no surface mixed layer is present, and if the sound speed decreases from surface to bottom, estimate the propagation loss as by bottom mode only, and take the propagation loss with two bounces as between 70 and 80dB.

The sound propagation in shallow water is subject to relatively great fluctuations. These fluctuations might have a wide spectrum of predominant periods. There is a definite seasonal period in these fluctuations, it being at its maximum during the winter. Other fluctuations have tidal periods, diurnal periods, and other shorter period fluctuations such as those caused by surface waves.

At present and in the near future, only the active sonar is and will be used in fisheries. In this sonar the environmental and man-made noises play a minor role.

In sound propagation the vertical gradients of the temperature are far more important than its absolute values. One is normally interested in the temperature gradient between the surface and the MLD to be able to know whether a surface channel exists. The MLD, in turn, determines the amount of energy which will be propagated in the surface duct as well as the dimensions of this duct. Finally, the gradient in and below the thermocline will largely determine the properties of the partial shadow zone.

The occurrence and magnitude of a near-surface transient thermocline has a considerable effect on the hull-mounted sonars. Moreovers, the various subthermocline ducts exercise a considerable influence on the sonar ranges below thermocline. Thus, knowing the thermal structure, a number of conclusions can be drawn on the sonar range and effectiveness.

Two sound devices are used for fish detection: the echo sounder and the sonar (or asdic). The beam of the echo sounder is directed downwards. Thus, only targets below the ship can be recorded. Both single targets and conglomerate targets, *eg* shoals of fish, can be recorded. However, the possibilities of discovering shoals of fish and single fish very close to the bottom are limited. The continuous traces do not allow for estimation of the abundance nor can any definite identification of the species be made, unless a catch be made, or the species is ascertained by other means such as underwater television. However, with experience and local knowledge one can make reasonable guesses and arrive at some expectations concerning the species which is being recorded.

The output of most echo sounders must be carefully controlled and periodically calibrated if one wants to record the abundance of fish quantitatively. The detection of fish near the bottom is improved when pulse length is kept as short as possible. Furthermore, higher frequencies tend to give a better resolution near the bottom. Under these conditions some measure of the size and density of a fish shoal can be obtained with echo sounders when the ship passes over them. If the amplifier is so adjusted that the noise is kept away from the records, one can count the traces at a unit distance and estimate the abundance from the results so obtained, provided some estimation of the calibration is possible.

No entirely reliable methods are available yet for direct identification of targets or for determination of the actual target strength. Some estimation of the latter is possible, provided the sonar or echo sounder has been properly calibrated. Therefore, in practical fishing a guess must be made about the species present, or an experimental catch must be executed. If the echo sounder has an especially narrow beam and uses a relatively high frequency, single targets in a less dense shoal may be counted sometimes.

One possible method for identification of the species, attempted by some workers, is to classify the traces of a calibrated sonar by their appearance and intensity. If then additional information on those types of traces are obtained, *eg* concerning the species, the density of fish in shoals, the dimensions of shoals, the location and the seasons, a 'calibration' of the recordings can be made which can then be used for interpretation of similar observations at a later stage.

Another hypothetical technique for identification of fish and determination of their size by modification of the frequency of the sounder or sonar, which has not yet been utilized, could be based on the resonance frequency of the air bladder. When having estimated the resonance frequency, the size of the air bladder of fish at a given depth can be ascertained, which in turn would allow an estimation of size and maybe of species of the fish present.

Another possible method for identification of the biological sonar targets has been described by Hester (1967) using the Doppler shift and the motion of the fish and boat. Dragesund and Olsen (1965) have described a method for the estimation of year-class strength of a fish species by measuring the echo abundance of juvenile fish. Here the main difficulty is the separation between the scattering groups and the scattering layers. The scattering layers of a groupy appearance – the so-called scattering groups – are discontinuous in horizontal plane and are only a few times larger in their horizontal than in their vertical extent. Thus, it could be assumed that high scattering cross-sections are typical for shoaling fish. Scatterers, which are continuously distributed in horizontal planes, form extensive layers the horizontal dimensions of which are very much greater than their vertical extent. Usually those layers appear on the record of an echo sounder as a uniform and weak echo trace.

In most sonars several automatic and manual search methods are available. In selecting a search method the sonar range for the given conditions must be known. This can be illustrated with a side-to-side search programme. Assuming that the sonar range was 500m and that we selected a 1 500m range for 15 knots speed, considerable pockets remain unsearched (*cf Fig 52*). From this single example, to which many similar ones could be added, it becomes clear that sonar predictions for fisheries are badly needed.

A number of different rules can be used in the search method with sonars with respect to the pulse length, the sensitivity control, and the beam width. Many of them depend on the special location, on the type of fish, and on the prevailing oceanographic conditions. Furthermore, experience with sonars is an important prerequisite for their successful use in fisheries.

Another important subject is the recognition and classification of different types of echoes, such as:

(*1*)   bottom echoes at different depths and from bottoms of different roughness,

(*2*)   side echoes in fiords,

(*3*)   echoes from secondary beams,

(*4*)   echoes due to refracted primary beams.

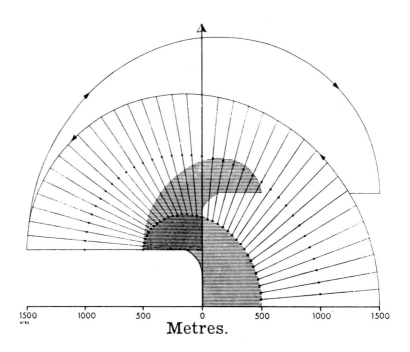

Fig 52   Sonar search programme side-to-side (after Vestnes, 1964).

108

Of primary importance is the recognition of different types of reverberation such as surface reverberation, bottom reverberation, and volume reverberation, as fisheries sonars often work in reverberation limited conditions. It should be pointed out that it is often difficult to differentiate between reverberation and scattered echoes from scattered herring shoals.

There are also static echoes from horizontal and vertical temperature discontinuities, which often are connected with biological scattering layers.

Two other types of echoes can be of considerable nuisance. First, there are the wake echoes, especially if many boats are working within a limited area. Secondly, there might be interference between the sonars and echo sounders of the same boat, as well as sonar interference from other boats working in the same area.

The tactical application of sonar depends greatly upon the type of fishery (eg midwater trawling or purse seining) and upon the depth and size of the fish shoals. Some hints from the Norwegian and Icelandic fishery tactics with sonars were described by Vestnes and Jakobsson (1964).

8

Ocean environment and its variability

The atmosphere and the ocean are a coupled energy system, with each providing some of the driving force to the other. The properties and the state of the interface between them (the sea surface) determine to a large extent the exchange of mass and energy between these two media.

It should be pointed out that synoptic changes in the surface layers of the sea, due to internal processes as well as exchange processes, are approximately on the same space and time scale as the processes in the lower atmosphere, and that there are in surface layers day-to-day changes of properties which can exceed the magnitude of annual range of monthly mean values of the same property, as will be shown later.

One may talk about advective changes, caused by currents, which in turn can be brought about by thermohaline (temperature and salinity) gradients, by winds, by waves (through the mass transport), by tides, by changes of atmospheric pressure and by other factors. Secondly, one may talk about local changes, usually caused by the heat and mass exchange between sea and atmosphere. Thirdly, one may consider the changes caused by mixing, which in turn can be forced by waves and currents or caused by convective stirring. There are other causes of changes as well, usually local in nature.

A given hydrographic situation does not repeat itself in the same manner from one year to another. Moreover, the surface circulation is not entirely bound to follow the local wind systems. Nevertheless, most of the causative forces for these changes are atmospheric in origin. Therefore, the horizontal extent of the changes is of an atmospheric scale, except in coastal regions where the effects are modified by the local topography. If we exclude the tidal changes and a few other types of sudden changes, we usually find that the average period (or cycle) of the changes is shorter (a few days) in higher latitudes, in the areas of passing cyclones, and is longer (about one week or more) in lower latitudes in the areas of the semipermanent anticyclones.

The amplitudes of the changes depend on the existing horizontal (and vertical) gradients of the properties in the surface layers of the oceans as well as on the driving forces in the atmosphere. For example, the local changes of the sea surface temperature can amount to 4°C in 12 hours near a sharp current boundary, but they are usually of the order of 0·3°C or even less over the major parts of the oceans.

110

8.1 Water properties and time and space scales

Summary

Monthly mean distributions of physical and chemical properties of the oceans are available in the atlases and in a great number of special (mostly local) reports. Thus the need for descriptive physical/chemical oceanography has diminished considerably in the last decade. Less information is available in literature on synoptic thermal structure with depth. Therefore, a classification of BT types and their description is presented in this chapter.

The submarine illumination is dependent, besides light intensity at the surface, on the turbidity of the water. Surface waters can be classified into optical water types and the submarine illumination can be predicted using these types.

Monthly mean current atlases are available for all ocean areas. The data from these atlases for estimation of surface currents are most reliable in areas of strong permanent current. The actual, highly variable, wind-driven surface currents can be estimated from wind data. Tidal current, which dominates on most continental shelves, can be predicted with hydrodynamical numerical models.

In the consideration and use of any distribution and change of environmental parameters, their space and time scales and the properties of parameters at the corresponding scale, must be considered. The conventional time scales are: climatological, medium-range, and synoptic; for which there are corresponding space scales: hemispheric, regional, and local scales. In fisheries work these 'natural' scales are complicated by 'administrative' restrictions.

The surface and near-surface water temperature and surface currents are those environmental parameters which are mostly used in deducing relations between the environment and fish behaviour and distribution. Both parameters are profoundly influenced by synoptic surface meteorological conditions.

There are numerous textbooks on oceanography which describe the oceanographic processes and distribution of properties. In this chapter we review only those oceanographic parameters which are important to fisheries ecology and which have not received adequate treatment in many textbooks, such as thermal structure with depth and light penetration.

The mean conditions in the oceans (*ie* the seasonal distribution of physical and chemical properties) have now been well investigated in most oceans. Atlases have been a common form for presentation of oceanographic data. Many atlases for surface layer parameters are available from national hydrographic services. The atlases of the deeper layer parameters of the ocean are rare and mostly old, such as the Atlases of the German *Meteor* Expedition. However, new atlases, where all available oceanographic data have been worked up via computerized processes, are now becoming available (*eg* Robinson and Bauer, 1976).

Unfortunately the numerical information contained in an older atlas is usually not comparable with the data given in others. The reasons for such discrepancies are numerous. First, different amounts of data, different numbers of years, and different ways of computations have been used to prepare the presentations. Secondly, different time periods have been used to derive the means for months, seasons, and even years. Even the methods of presentation vary considerably. Most of the parameters presented refer to the surface conditions, such as waves, surface temperature, and surface weather conditions. In the case of waves, different statistical charts with the percentage of seas higher than a given value are usually presented. Another popular subject for climatological atlases has been the presentation of surface currents, usually derived from logbooks, based on dead reckoning.

Quasi-synoptic oceanographic data have been presented also as sections and/or charts of relatively small areas. The observations of one or a few research vessels have been used for these presentations. Such sections and charts can be found in numerous monographs and articles on descriptive oceanography.

Oceanographic station data are often reproduced in data lists such as the *Bulletin Hydrographique* of the International Council for the Exploration of the Sea and the monographic work of *Oceanographic Observations of the Pacific*. As a result of the rapid increase in the amount of oceanographic data, and of the availability of electronic processing, the data are currently kept on punched cards or magnetic tapes.

Schematic presentation of thermal structure with depth and its seasonal changes is given in *Fig 53*. Further definitions of additional features in thermal structure with depth are given in *Fig 54*.

The mixed layer depth (MLD) is defined as the thickness of the turbulent, homogenous surface layer. However, when this simple definition is applied to the complex thermal structures of the ocean many ambiguous cases are found (see *Fig 55*). There are three main types of bathythermograph (BT) traces with the ambiguity of this definition:

(*1*)   where the temperature is nearly homothermal to great depths,

(*2*)   during the early stages of the formation of a seasonal thermocline in areas where a permanent thermocline occurs, and

(*3*)   where strong currents exist.

A nearly homothermal BT trace from the surface to a relatively great depth is observed in some higher latitude areas, and during the winter in the Mediterranean. When at times the mixed layer depth (MLD) can be verified by small discontinuities in the temperature trace, it may be called the potential MLD (Tully and Giovando, 1963).

During the spring in most ocean areas a seasonal thermocline starts to form in the layer near the surface. When this occurs in an area with a permanent thermocline either the top of the permanent thermocline or the depth of the seasonal thermocline (or any

112

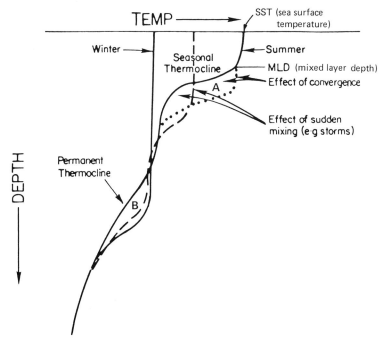

Fig 53 Schematic diagram of thermal structure with depth indicating effects of conditions and processes in the water column. Greatest temporal changes take place in regions marked with A and B.

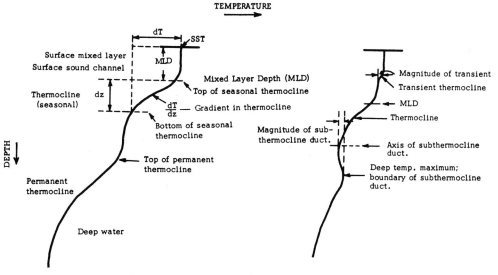

Fig 54 Some general definitions of the near-surface thermal structure and examples of parameterization of a BT trace.

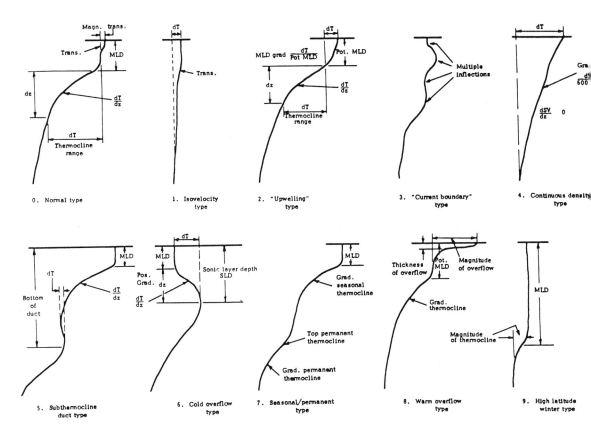

Fig 55 Types of BT traces and their interpretation.

transient thermocline) can be considered the MLD. However, as soon as the seasonal thermocline has a magnitude of approximately 1°C, its depth is considered the MLD.

In areas of strong currents, especially at sharp current boundaries, the thermal structure becomes quite complicated and the simple description of a mixed layer above a smooth thermocline layer is no longer accurate.

The surface layer is usually homothermal in autumn and winter (when the sea is losing heat to the atmosphere), except in areas near the current boundaries. In spring and summer (when the sea is gaining heat) the surface layers often are not homothermal but contain small transient thermoclines. Depending upon the purpose for which the information on the MLD is applied, other definitions of the MLD might be used in some cases. For climatological purposes the MLD is often defined as the depth at which the temperature differs 1°C from that at the surface. Obviously this difference is in most cases negative. For sonar purposes the mixed layer depth is often defined as the surface sound channel above the sonic layer depth (SLD) which is the depth of maximum sound velocity. In the absence of a sonic layer depth, the surface sound channel is defined as

114

the region with isovelocity profile (or with a sound speed profile showing an increase with depth). In still other cases, the MLD is assumed to reach the top of the seasonal thermocline or any other significant temperature, or salinity, discontinuity near the surface.

Figure 56 summarizes the major factors determining the mixed layer depth and bringing about its fluctuations. The first major cause of a well-mixed surface layer is the mixing by wave action, both by the particle movement by waves as well as by the turbulence caused by breaking waves, especially by winds strong enough to generate wave heights larger than 3m. The depth of the mixing by waves is influenced by the stability of the thermocline (*ie* by the sharpness of the thermocline and by the connected salinity gradient).

The second major cause for a change in the MLD is the heat exchange. A positive heat exchange means the gain of heat by the sea creating, among others, the seasonal thermocline during spring and summer. On the other hand, a loss of heat during autumn and winter causes a convective turnover, thus creating a well-mixed surface layer and deepening the MLD.

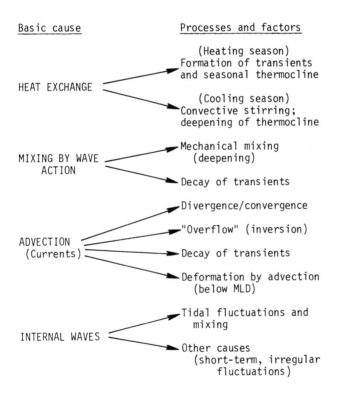

Fig 56 Factors determining the mixed layer depth (MLD) and its changes.

In areas with strong currents, the thermal structure can be complex indeed and is greatly influenced by different currents from differing directions at different depths. Short-term fluctuations of the thermocline depth are affected by the so-called internal seiches and tides, by the divergence/convergence, and by various other causes. It should be noted that the whole thermal structure moves up and down with divergence and convergence with the vertical gradients more or less unchanged but with the mixed layer depth changing (Patullo and Cochrane, 1951; Robinson, 1966).

The frequency distribution of the mixed layer depth is usually normal in a given location. Exceptions to this rule are the areas near a water type boundary or near a current edge and near the continental shelves. An example of the frequency distribution of the MLD is shown in *Fig 57*. If a larger area is used for such a statistical treatment, the frequency distribution is 'flattened' by the existing horizontal gradients of the mixed layer depth.

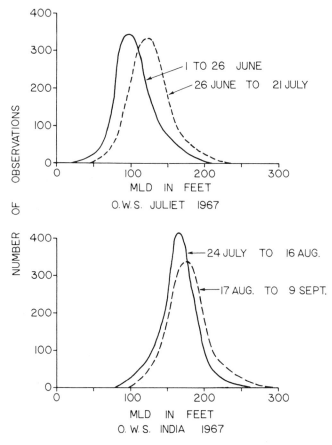

Fig 57 The frequency distribution of the MLD based on BT casts every 15 minutes (after Wood, *pers comm.*).

116

The thermal structure obtained with a mechanical BT can be relative. It is still, however, a useful tool for obtaining the general thermal structure and for ascertaining the small-scale space and time variability in the surface layers.

The general information obtainable from a BT trace is illustrated in *Fig 54*. It is obvious that for many applications the sea surface temperature (SST) and the mixed layer depth (MLD) data are insufficient. Consequently the whole temperature profile must be known. It is especially important to know the whole temperature profile if the gradients in the thermocline and the gradient between the MLD and the surface are to be applied to sonar problems.

The temperature values obtained with a mechanical BT are not always absolute values because of the drift in calibration of the instrument. The surface temperature of a BT might also be incorrect, at times, due to a short acclimatization of the BT in the surface layer.

A number of different thermal structure profiles are given as *Fig 55*. Some brief characteristics of the more frequently found bathythermograph profiles (BTP types) are given below:

The normal (or standard) BTP type (O) consists of a mixed layer above a thermocline and deep water below it. The mixed layer may contain transient thermoclines, especially during the spring and early summer. During the autumn, the mixed layer is usually homothermal due to convective stirring, caused by heat loss (or by a higher evaporation) at the surface.

The seasonal permanent type (7) is in its uppermost part similar to the normal type but contains another, permanent thermocline at a deeper layer. During the spring in the medium and high latitudes, a seasonal thermocline starts to form as a transient thermocline caused by warming at the surface. The depth of this transient is usually variable and determined by the prevailing wave conditions. The mixed layer depth slowly deepens during the course of the summer and starts to descend more rapidly in October and November, when cooling occurs at the surface. It reaches the permanent thermocline usually between January and March. Thereafter this BTP type might be considered as a normal type during the rest of the winter. Thus, in some areas there may be a change from one BTP type to another during the course of the year. This seasonal cycle does not occur in the low latitudes nor in areas of strong currents, nor in upwelling areas.

The upwelling type (2). In the areas of upwelling, the mixed layer depth is usually shallow. The temperature gradient between the top of the thermocline, which at times is ill defined, and the surface can be considerable. The thermocline range and gradient can, at times, be sharp indeed, especially at the continental slopes where the main upwelling occurs.

The continuous density type (4) often occurs both in upwelling areas and also during the spring and summer in high latitude areas.

The subthermocline duct type (5) occurs in relatively large areas in the North

117

Pacific and in similar areas in the North Atlantic. In some parts of the Sargasso Sea and northeast of the Japanese Islands this thermocline type might also appear but in a less pronounced form. It is characterized by a slight temperature inversion below the seasonal thermocline. The inversion layer usually contains water of a slightly lower salinity.

The high latitude winter type (9). The medium and high latitudes in the North Atlantic are characterized by a very deep mixed layer depth and a weak thermocline at a great depth. In still higher latitudes this small thermocline might be absent, resulting in the isovelocity type (1).

The current boundary type (3) is found at the boundaries of major currents such as at the Kuroshio-Oyashio and the Gulf Stream–Labrador Currents. It can contain a very thin mixed layer and a multiple of inflections and temperature inversions. These inversions and inflections are rapidly variable in space and time.

The cold overflow type (6) is created by colder, less saline water on top of a warmer, more saline water. This type occurs during the winter in the Gulf of Maine and the St. Lawrence Gulf.

The warm overflow type (8) occurs when relatively warm water is advected over cold water. It occurs near current boundaries and is usually of short duration.

It is often necessary to determine the bottom of the thermocline. This determination is rather ambiguous but can be accomplished by extending graphically on the BT trace the sloping thermocline downwards and the temperature trend line of the deeper waters upwards and projecting the intersection of these straight lines horizontally to the original temperature traces.

Temperature changes in the deeper layers are much smaller and slower than above the thermocline although an exception is the layer near the thermocline(s) where relatively sudden changes can occur due to mixing and/or convergence (see *Fig 53*, locations A and B). In fact, sudden changes in the depth of the surface layer and, thus, the thermocline, might act as triggering mechanisms or stimulus for the initiation of seasonal migrations of various fish.

The light intensity at any depth in the water is determined by a multitude of factors. First, it is a function of light intensity at the surface, which in turn is a function of latitude, time of the year, time of the day, cloud cover and type, and general weather condition (*eg* rain, fog, haze, *etc*). The light penetration in the water is determined by the turbidity of the water and by the wave-length of light. The water turbidity is used to classify the surface waters into optical water masses.

For the estimation of the extinction of light, the curves for incident daylight, given in *Fig 58* serve as examples. These curves have been constructed for the five different optical water masses, defined in *Table 7* for which the spectral extinction coefficients are given in *Fig 59*. A more detailed breakdown is probably unnecessary for all practical problems.

The climatological surface current charts are widely used to estimate the current

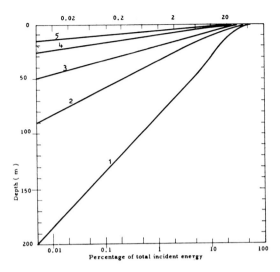

Fig 58 Percentage of total incident daylight energy at different depths in different optical water masses (modified from Jerlov and Kullenberg, 1946, and Jerlov, 1951).

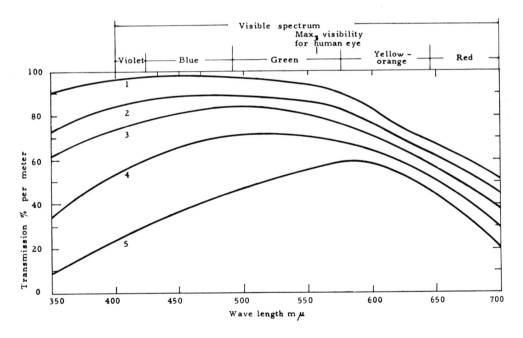

Fig 59 Average spectral transmission curves for different optical water masses (from Jerlov, 1951, and Jerlov and Kullenberg, 1946, modified).

119

Table 7
Optical water masses

No.	Water mass	Characteristics
1.	Oceanic, clear	'Old' clear oceanic waters in low-productive areas (especially in low latitudes). Water colour 0 to 2 (Forel Scale). (J: Oceanic 1.)
2.	Oceanic, normal	Medium-productive oceanic water in medium and low latitudes. Water colour 2 to 5. (J: Oceanic 2 and 3.)
3.	Oceanic, turbid and Coastal, clear	High-productive oceanic areas, especially during plankton bloom. Tropical coastal waters, especially over deep shelves. Water colour 5 to 8. (J, K: Coastal 3.)
4.	Coastal, normal	Normal, medium-productive coastal waters and waters over shallow shelves. Water colour 8 to 10. (J, K: Coastal 5.)
5.	Coastal, turbid	Estuarine and coastal waters during intensive plankton bloom and waters close to the coast where much sediment has been whirled up by wave action. Water colour 10. (J, K: Coastal 9.)

Abbreviations: J, K – Jerlov and Kullenberg, 1946. J – Jerlov, 1951.

drift. These charts, constructed from data on ships' drift, are available for nearly all ocean areas. The reports upon which these atlases are based were obtained from logbooks and are nothing more than the difference between dead-reckoning fixes (from ships' speed and direction) and navigational fixes (astronomical, Loran, visual, *etc*). Their accuracy and usefulness have, however, severe limitations. The data presented in the current atlases are useful as a rough estimate of the actual current at a given time and place.

Hela (1954) noted that if the current displacement, determined by the astronomical fix, is less than 2 nautical miles per 24 hours, the drift value in an atlas is entirely unreliable. In most cases, little weight can be given to displacements of 5 nautical miles or less per 24 hours. James (1957) came to an analogous conclusion: the reports on the surface current charts can be considered reliable and accurate as mean values if there are at least five observations from 3 600 square nautical miles and if the drift is more than 5 miles per 24 hours. There is a real need for synoptic current charts, particularly for offshore areas.

Numerous detailed current measurements, taken with instruments of various types, are now available for limited areas. These have been useful in building up our knowledge and theories on the surface currents. Most of these measurements have been made relatively near the coast since the measurements of actual surface currents offshore are extremely difficult to make. Thus the coastal and tidal effects limit the value of current measurements when drawing conclusions on the offshore currents. Many of the current measurement series are also too short for a proper evaluation.

The currents in the surface layers can be classified into three main categories by the driving forces:

(*1*) Tidal current, which is the prevailing current component in most shallow water over the continental shelf. Tidal current is not absent over deep water, but is considerably weaker there. Tidal currents are predicted numerically, using Hydrodynamical Numerical (HN) models.

(2) The wind-driven current component varies with the surface wind (with some time lag) and is computed from surface wind data (see Chapter 10).

(3) The 'permanent' or thermohaline current component is a large-scale current component caused by thermohaline pressure gradients and includes also large-scale wind system component. This current component is usually obtained from existing large-scale ocean current charts.

It is necessary for better understanding, description, and use to systematize the changes in the environment by time and space scales. There are three basic time scales: climatological (including year-to-year and seasonal time summaries), medium range (few days to two to three months), and synoptic scales (few minutes to few days).

(a) The climatological means

Many climatological means, such as monthly means, seasonal means, and annual means, of the distribution of a given property, *eg* of the sea surface temperature, are possible. The most commonly used are the monthly means in the form of charts or tables. There can be considerable differences between the presentations of monthly means, prepared in different laboratories, depending on the amount of data available and on the methods used. Pronounced secular changes in some properties of the ocean, *eg* in the sea surface temperature, have been clearly demonstrated by a number of workers. The monthly means are often used as a basis for computation of the anomaly for given time periods. Thus apparently different anomalies result from the use of different climatological means.

The seasonal and monthly forecasts of some properties are used in some fisheries problems. Basically they are extrapolations from the long-term trends with due attention also to the anomalies and persistency of the trends.

(b) Medium-range forecasts

The medium-range forecasts (mainly of surface meteorological parameters) (from five to ten days) are mostly needed in marine operations such as in fisheries and ship routing. Attempts to achieve such forecasts have been made in many countries, and five to ten day outlooks of ocean areas are now available. They are usually based on the extrapolation of the movement of existing systems, on the use of analogies, on the climatology, and to a limited extent also on the possibilities of predicting a cyclogenesis in defined areas and conditions. Many attempts have been made to find a physical explanation of the success of these forecasts such as the feedback from the ocean and the importance of the sea surface temperature anomalies in this feedback. Most of the earlier attempts have fallen short of the scientific basis required. Owing to the considerable economical pressure and to the acute need of such forecasts more serious work is in progress in many countries and institutions for the improvement of such forecasts. This would most naturally come from the extension of the present short-term synoptic weather forecasts which have a physical and usually also a mathematical basis.

(c) The synoptic time scale

Most of the existing weather analyses and forecasts fall within the synoptic time scale which ranges in most cases from 12 hours to about three days. The synoptic oceanographic analyses and forecasts started in the last decade fall within the same time scale.

Every process in nature has also space scales of distribution and variability. Some of the space scales find correspondence in time scales. Examples of area and time scales of sea surface temperature changes are given in *Table 8*.

(a) *The hemispheric scale.* The hemispheric space scale is needed in climatological as well as in synoptic work. The 500 millibar analyses in the atmosphere, with its constantly changing waves, and also with the large-scale interactions between the cyclones and anticyclones can serve as an example. Hemispheric scale analyses will also serve for defining proper boundary conditions for analyses of smaller scale.

(b) *Regional scales.* The regional scale cannot be defined implicitly. However, one quarter of a hemispheric half of an ocean is usually considered to be a region in the meteorological sense. A more natural division is by the so-called natural oceanographic regions, where the boundaries are defined by continents, by ocean currents, and by convergence and divergence zones (see *Fig 60*).

(c) *Local and small-scale analyses.* A 'local scale' analysis of distribution could be made for example for the North Sea, or for a smaller natural region, whereas a 'small-scale' analysis would be made for a small bay or for an estuarine area. The scale of the analysis determines to a large extent the density of observations required and the accuracy of the analysis concerned. The scales must also be selected in an appropriate relation to the time and space scales of the motions. The small-scale and local analyses require also much better local knowledge and experience than the regional analyses.

Besides the 'natural' time and space scales, there are space scales which are administratively defined. The administrative scales of marine and fisheries services are to a certain degree determined by the space scales of ocean regions and by the distribution of fish communities which determine the types of fisheries. First, in administrative scales we have the international organizations with global interest. Theoretically, these organizations should be dealing with the hemispheric scales as well as with the long-term time scales (climatologies and long-term economic problems). However, in practice this has become unworkable because of the bureaucracy and politics involved. In practice, one or a few nations are responsible for work in the hemispheric scales of environmental analyses/forecasting. More efficient are the regional organizations, inter-governmental by nature, such as the International Council for the Exploration of the Sea, Inter-American Tropical Tuna Commission, and others. The local administrations are usually concerned with one or a few particular fisheries. Either a regional laboratory in a given state or a fraction of a state laboratory is normally responsible for a given fishery or region.

Table 8

Approximate area and time scales of the SST changes

Processes	Approximate area scales	Approximate time scales
1.1. Permanent (gradient) flow	Usually in oceanwide scale and in form of gyrals, 500 to several thousands of km in diameter. Small off estuaries and modified near continental shelf.	Seasonal, except near current boundaries and, in coastal waters where dependent on insolation and runoff.
1.2. Wind currents	Gyrals correspond to the sizes of cyclones and anticyclones.	Cyclone belt – 2 to 8 days. Anticyclone belt – 6 to 14 days.
1.3. Inertia and tidal currents	Size of the amphidromic tidal systems. Smaller in semi-closed bays.	Tidal; diurnal or semidiurnal. Inertia currents dependent on latitudes (av. 30 hrs.).
2.1. Insolation	Greatly determined by latitude and by cloudiness patterns. In general one half of a cyclone and one fourth of an anticyclone size. More rapidly changing smaller patterns in tropical storms, at coasts, and occasionally at sharp current boundaries.	Seasonal and synoptic (see 1.2 above).
2.2. Evaporation		Mainly seasonal; synoptic periods as 1.2 above.
2.3. Other heat exchange components		Seasonal and synoptic as 1.2 above. The synoptic periods also vary seasonally, especially at low latitudes.
3.1. Wave action	Generally the size of cyclones and their wind fields.	Cyclone belt – 2 to 8 days. Anticyclone belt – 6 to 14 days.
3.2. Convective stirring	Generally latitudinal pattern, at the periphery of cyclones (about half of their size).	Mainly seasonal, in medium and high latitudes.
3.3. Currents (mixing by)	Usually important near major current boundaries. Scale from a few miles to a few hundred miles.	Seasonal, except near current boundaries and in coastal waters.
4.1. Upwelling and divergence/convergence	Usually narrow and elongated areas near coasts, oceanic or atmospheric fronts; from tens to hundreds of miles wide, several hundreds to thousands of miles along.	Seasonal and synoptic (see 3.1 above).
4.2. Runoff	Off estuaries and along the coasts; a few miles to a few hundreds of miles wide.	Mainly seasonal.
4.3. Precipitation	Of some importance only in high latitudes during the winter; the size of precipitation (snow) area.	Seasonal and synoptic (see 3.1 above).
4.4. Freezing and melting of ice	Important only near coasts and near ice boundaries.	Mainly seasonal.

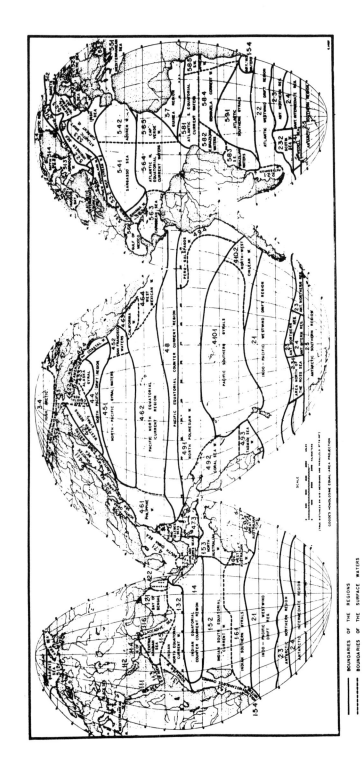

BOUNDARIES OF THE REGIONS
BOUNDARIES OF THE SURFACE WATERS

Fig 60 Natural regions of the oceans.

124

8.2 Temporal changes and ocean anomalies

Summary

A number of factors and processes can act simultaneously to change sea surface temperature (SST) at a given location. The SST change at a given location during a few days can reach and exceed half of the mean annual range of SST change at these locations.

The variability of oceanographic parameters is considerably larger in coastal than in offshore areas. Furthermore, small-scale anomalies change faster than large-scale anomalies.

The boundaries of permanent current systems change seasonally and there are small synoptic changes in the positions of these boundaries. The change of oceanographic parameters at different sides of the oceanic fronts can have trends with different signs.

The surface wind-driven currents, as well as upwellings, change in the same rhythm as the surface wind.

It is doubtful whether fish can respond to short-term (synoptic) changes of environment in horizontal direction except by onshore-offshore movement. The main synoptic response of fish is in vertical direction (*ie* within depth).

In order to evaluate the effects of environment on the fish biota, it is necessary to know the magnitudes of temporal changes in the oceans, including the periods of such changes and their spatial scales. Quantitative knowledge of the short-term changes in the ocean is needed also for drawing conclusions on anomalies or changes in the ocean from single monthly or seasonal surveys as short-term variations may mask the seasonal variations. This can be illustrated as follows: the sea surface temperature (SST) and the mixed layer depth (MLD) can change within a few days more than 1°C and 30 metres respectively, especially if a storm passes over an area. If a 'monthly' or 'seasonal' survey is made in the area within a few days or a week, the measured conditions do not represent the mean for the section or area, but rather the synoptic, changing conditions.

One of the main objectives of this chapter is to indicate the amplitudes and the average periods of short-term variations in the surface layers, and to demonstrate their causes. The observational errors are closely related to the problems of short-term variability; however, they will not be referred to here.

The factors affecting the sea surface temperature (SST) and the resulting 48 hour changes are given in *Table 9*. (*Table 8* indicates the approximate area scales and periods of these changes.) There are four major causes for the SST changes; advection by currents, heat exchange between sea and atmosphere, mixing by wave action, and convective stirring.

Mixing should be considered as occurring mainly in vertical direction where the temperature gradients are large. In some areas upwelling and divergence/convergence can cause changes in the SST. There are a number of minor causes for the SST changes

125

which are of limited extent, such as the runoff from river estuaries, precipitation in the form of snow or hail at high latitudes, and freezing and melting of ice in very high latitudes especially in coastal waters. The rate of the SST change, due to a single process, may not be representative of the total changes occurring; *eg*, an intensive local warming may result from the advection of warm water but at the same time a compensating heat loss may occur. Thus the existence of a positive anomaly of the SST does not necessarily indicate whether there is a minimum of heat removal or a maximum of heat supply, and *vice versa*. All factors affecting the SST must be known and be taken into consideration simultaneously in order to establish a valid budget for a given location and time.

Column 4 in *Table 9* lists the assumed values on which the possible SST changes in 48 hours are based. As *Table 9* indicates, a temperature change of 0·3 to 1·7°C (0·5 to 3°F) in two days would be possible in many areas of the oceans. However, there exist even more favourable conditions for large SST changes than those assumed in these calculations. In various areas the magnitude of the surface temperature change during a short period of a few days may equal the total annual range. This is especially true in medium and lower latitudes and in cases of upwelling (see Rodewald, 1964). The greatest change obviously is achieved if several factors affect the SST in the same direction. Since the main factors bringing about cooling are mixing and convective stirring, cooling can be more rapid than warming. In areas with a relatively shallow mixed layer, a change of 2·8°C (5°F) in 48 hours is observed rather frequently. Column 2 in *Table 8* indicates the area scales of the changes in the sea surface temperature, as caused by different processes. The scale of the SST changes usually corresponds to the scale of the driving forces; however, smaller scale changes are possible in certain situations. If there are several processes occurring simultaneously, an overlapping of several scales is probable. A moving atmospheric system changes the scale of an SST change. Scale changes take place also in coastal waters and around islands as a result of topographic and bathymetric effects.

The last column in *Table 8* gives approximate time scales of the SST changes caused by different processes. Most of the processes are aperiodic and can be described only by average or mean periods. In reference to the meteorological driving forces, such as wind currents, heat exchange, and waves, one can classify the periods of change into short and long periods. Short period changes, of two to eight days, occur in the cyclone belt at higher latitudes, the periods being longer during summer and shorter during the early part of winter. Six to fourteen days occur in the anticyclone belt of the lower latitudes. These periods also vary with the seasons.

The SST changes during a few days only can be of the order of half the annual range of the monthly mean sea surface temperatures. In some instances they even exceed the total annual range. The actual, truly diurnal variations of the SST are influenced by the heat exchange factors. These changes are, however, relatively minor as compared with the possible errors of measurement and reporting. *Table 10* gives the estimated average values of diurnal changes of the SST in various conditions.

Table 9

Factors affecting the sea surface temperature (SST) and the resulting 48 hour changes (Wolff, 1967)

Basic cause	Contributing processes	Assumed values for computation of change in 48 hrs.	Possible SST change in 48 hrs. °C	(°F)
1. Advection	1.1. Permanent (gradient) flow	Speed 1 knot; SST gradient 1·7°C (3°F)/ 100n miles	0·8	(1·4)
	1.2. Wind currents	Speed 0·4 knots, SST gradient 1·7°C (3°F)/ 100n miles	0·2	(0·4)
	1.3. Inertia and tidal currents	Speed 0·4 knots, SST gradient 3·3°C (6°F)/ 100n miles	0·4	(0·7)
2. Heat exchange	2.1. Insolation (affected by clouds)	600g cal cm^{-2} 24 hrs^{-1} MLD* 17 m (50 ft)	0·8	(1·4)
	2.2. Evaporation (affected by wind and $e_w - e_a$)	300g cal cm^{-2} 24 hrs^{-1} MLD 17 m (50 ft)	0·4	(0·7)
	2.3. Other heat exchange components	200g cal cm^{-2} 24 hrs^{-1} MLD 17 m (50 ft)	0·3	(0·5)
3. Mixing	3.1. Wave action	Deepening of shallow MLD by 8·5m (25 feet) with sharp gradient below	1·7	(3·0)
	3.2. Convective stirring	Dependent on heat loss	0·6	(1·0)
	3.3. Currents	Dependent on sharpness of boundaries	0·17	(0·3)
4. Special causes	4.1. Upwelling and divergence/convergence	Gradient 2·8°C/30 m (5°F/100 ft), Uplift of lower limit of MLD resulting from divergence 15 m (50 ft)).	(1·4)	((2·5))
	4.2. Runoff	Important off estuaries	(0·3)	((0·5))
	4.3. Precipitation	(Important only in case of snow and hail)	(0·1)	((0·2))
	4.4. Freezing and melting of ice	(Important in limited areas in high latitudes)	(1·7)	((3·0))

*Mixed Layer Depth

Table 10

Estimated average magnitudes of diurnal variations of the SST (°C) in summer and winter at various latitudes in offshore areas. (Based on numerous published sources.)

Latitudinal zone	Summer (Apr-Sept) shallow thermocline Wind force (Beaufort)					
	0–2		3–5		>6	
	Clear	Cloudy	Clear	Cloudy	Clear	Cloudy
0°–20°	1·5	0·8	0·6	0·3	0·2	0·1
20°–40°	1·0	0·3	0·4	0·2	0·1	0
40°–60°	0·5	0·1	0·2	0·1	0	0
>60°	0·2	0	0·1	0	0	0

Latitudinal zone	Winter (Oct-March) deep thermocline Wind force (Beaufort)					
	0–2		3–5		>6	
	Clear	Cloudy	Clear	Cloudy	Clear	Cloudy
0°–20°	1·2	0·5	0·3	0·1	0·1	0
20°–40°	0·5	0·1	0·1	0	0	0
40°–60°	0·1	0	0	0	0	0
>60°	0	0	0	0	0	0

The actual interdiurnal changes may be of the same order of magnitude as the regular diurnal changes. They are frequently considerably larger, especially near the major convergences of currents and in areas where advectional effects are large.

A number of synoptic sea surface temperature changes are presented in various forms in the available literature. Sagalovskii (1958) showed that the average day-to-day change of the SST in 24 hours at the Atlantic weather ships was 1·03 to 1·26°C in February, and 0·93 to 1·22°C in August. The corresponding diurnal change was 0·33 to 0·74°C in February and 0·33°C in August.

Considering the speed of the eddy diffusion processes in the ocean, in relation to the factors affecting the changes of the sea surface temperature, it becomes obvious that in the surface layers there must be a certain amount of 'thermal noise', the magnitude of which varies from season to season and from area to area. It is relatively difficult to find numerical parameters for this thermal noise as instrumental and observational errors are superimposed on it.

The variability in the ocean is slightly higher in the coastal waters than in the offshore water. There are also long period (secular) changes in the ocean which have some influence on the distribution and abundance of some fish, especially near the northern boundaries of their natural distribution.

The importance of the large-scale anomalies of the SST over the North Atlantic on the climatic fluctuations was investigated and described by Helland-Hansen and Nansen (1920) (see further Chapters 8 §8.3 and 9). The small-scale anomalies are the subject of continuous interest by fisheries scientists, especially in the International Council for the Exploration of the Sea.

A few general observations on the behaviour of small-scale anomalies are given below. The varying degrees of variability of these anomalies bears out earlier observations that there are ocean areas which are in relative thermal rest and other areas in thermal unrest. Several coastal anomalies indicate upwelling and its intensity (*eg* off San Francisco). These respond to the prevailing wind systems. The advective anomalies at major current boundaries (*eg* Labrador–Gulf Stream) are persistent, but show changes according to the prevailing local winds. This was already noticed by Chase (1959), who found that advection was an important factor in changing the local SST at Frying Pan Shoals. In general, warming occurred ahead of cold fronts (when southwesterly winds are prevalent) and cooling occurred after the frontal passage (when northerly winds prevail).

It can be noted that advective changes on the opposite sides of a major current boundary may have opposite trends. This has been pointed out also by Templeman (1964) who indicated that neighbouring waters with differing origins and temperatures may exhibit different temperature trends.

The use of the SST anomalies in fisheries has been demonstrated during the past half century. A typical description of the great negative anomalies in the North European waters, in winter 1962/63, and their influence on fisheries, has been given by Eggvin (1963).

The SST anomalies have proved to be useful in subjective analysis of the upwelling intensities of the surface eddies (especially of the warmer eddies in a colder environment), in delineation of the major current convergences and their fluctuations and in estimation of the subsurface thermal structure. Further use of the SST anomalies can be and has been made in the study of causal relations of the sea-air interaction and in the extended weather forecast for ocean areas. The last-mentioned aspect has been explored and described by Namias (1963).

The variability of the currents is considerable but attempts to evaluate quantitatively the causes of these variations have been few. This can best be done by means of the forecasting and daily verification. The forecasts are of little value only if they cannot be verified by actual synoptic analyses. Owing to the recent development in other fields of oceanographic analysis and prediction, both direct and indirect verifications are now possible. Through these verifications, additional knowledge is gained and existing ideas refined.

The boundaries of major currents – *ie* the current convergences and divergences, create major environmental boundaries (fronts) in the oceans.

The environmental fronts in the oceans are in principle comparable to the atmospheric fronts: they delineate the boundaries of surface water types with different physical-chemical and biological properties. The oceanic fronts, however, have a greater variety than the atmospheric ones. The sharpness of oceanic fronts can range from a well-defined demarcation to undefinable transition regions which still constitute a dynamic front (current boundary). Some fronts are stable in space and time, some move considerably and change in intensity.

The knowledge of positions, nature, intensity, and dynamics of the fronts has a multitude of applications in fisheries, in naval problems, and even in merchant navigation.

The fronts can be recognized in many instances by instrumental as well as by visual observations. The usual indicators of fronts are rapid changes in sea surface temperature and water colour, modified surface waves, accumulation of debris, and sea smoke.

Numerous unsummarized observations of ocean fronts are available in ships' logs. Inspired by the enthusiasm of Maury, one of the first international conferences on meteorological matters, held in Brussels in 1854, agreed to observe and to report voluntarily not only meteorological elements but also any other natural phenomena which are encountered during the voyage. A wealth of observations of various types of discontinuities in the sea surface temperature and water colour and other frontal phenomena has accumulated over the past one hundred odd years. Some of the interesting observations by the British ships are reported in *The Marine Observer*. Unfortunately, no extensive working-up of these frontal observations has been done in the past. The best earlier climatological frontal charts appear to be those of Schott (1931, 1942).

Griffiths (1965) discussed the definition of fronts in oceanography and found that no precise definition is available, nor practical. He accepts LaFond's (1961) definition and description of the oceanic front: 'The leading edge of a border separating two water masses, different in their properties, is called a front. Fronts can occur not only between water masses of different salinity but also between those differing in other properties, such as temperature.'

The nature of the oceanic fronts can be quite variable. LaFond (1961, 1963) and Griffiths (1965) gave some physical and biological descriptions of fronts. More generalized frontal models with indication of subsurface thermal structure at fronts were given by Hela and Laevastu (1962).

Oceanic fronts are most often boundaries between different current systems (divergences/convergences) and thus also boundaries of different surface water types. Hence, they form boundaries between different natural regions of the oceans.

In the past some attempts have been made to divide oceans into natural regions, more or less corresponding to the climatic regions of the land areas. The earlier divisions were usually made on the basis of a few characteristics only. An attempt to revise these divisions on the basis of combined environmental (including biological) characteristics is given in *Fig 60*.

The knowledge of the characteristics and environmental properties of these regions finds considerable application in a subjective description of the oceans and also in subjective oceanographic forecasting.

The seasonal changes of most surface environmental parameters follow a regular pattern over large areas. Seasonal changes below the surface and specially near the bottom, the main domain for demersal fish, are usually considerably smaller in magnitude and lag behind the changes at surface. The seasonal changes of thermal structure with depth and the delay in warming of the deeper layers are schematically shown (*Fig 61*). This warming does not necessarily follow a smooth, seasonal pattern, but might occur within a day or two when the first winter storm passes through a given area. Thus, some seasonal changes may be quite abrupt, and their timing can vary considerably from year to year. These changes appear to have a great influence on fish and their behaviour by triggering seasonal migrations, which are known to begin rather abruptly in many species.

Biological changes in the sea (*eg*, abundance of a species, or start of a plankton bloom, *etc*), as well as changes of environmental parameters in subsurface layer, follow a stepwise pattern; the decline being relatively sharp, the recovery slow. This condition does not apply to the changes of environmental parameters at or near the sea surface, which are caused by advective processes. The pronounced seasonal changes are also the main causes of various environmental anomalies that, once established during a transition season, spring or autumn, usually persist through the following season, summer or winter.

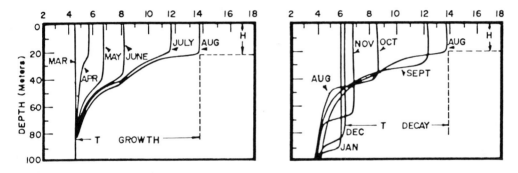

Fig 61 Seasonal types of vertical thermal structures and growth and decay of the thermocline at Ocean Station 'P' in the Gulf of Alaska (after Dodimead *et al*, 1963).

The surface wind drift (Ekman transport), or flow in the surface layer of the ocean, is driven by the direct action of the wind (the net effect occurs at right angles to the wind – to the right looking downwind, in the northern hemisphere) and has a marked effect on not only the drift of eggs and larvae, but also movements of juvenile and adult fish. Because of its limited vertical extent, this flow is always almost negligible in terms of total transport in the water column, but it results in very large convergence and divergence patterns. Computer techniques permit compilations of data on a daily basis that are assimilated into longer time periods of general analyses. Although winter intensification in wind-stress dominates the annual cycle of this flow, the seasonal shifts in winds result in various combinations of inshore and offshore convergences and divergences near the coast because of the physical barrier imposed by the land. When high southerly winds occur near a north-south oriented coastline, the increased shoreward Ekman transport piles up water along the coast that sinks (downwelling), and the seaward flow at depth satisfies, in part, the local increased requirement for surface flow. When the offshore winds are more intense than those at the coast, sinking occurs along the coast and offshore as well. The reverse happens when north winds occur; when the north winds offshore are more intense than those inshore, surface flow both at the coast and offshore has a seaward component and local continuity is satisfied by a vertically upward movement (upwelling) from depth, in both areas. And, finally, when northerly winds are most intense at the coast, there is an intense seaward flow at the coast that diminishes offshore and results in upwelling both on the coast and offshore.

A high rate of change of environmental parameters can in some cases affect a biological subject more profoundly than a large, slow rate of change; the former often causing shock effects, whereas the latter allows some acclimatization. Thus, sudden changes, as well as abrupt environmental gradients, are known to trigger behavioural changes (migrations, changes in feeding habits, *etc*) in fish which inevitably affect their abundance and availability. Short-term changes in environmental parameters in the sea are caused by passing atmospheric systems and are caused mostly by advective and mixing processes. An example of an advective type, short-term change of sea surface temperature at weather ship *November* is presented (see A *Fig 62*). If the readjustment by a driving force is not operative, an anomaly, caused by a short-term change, can be rather persistent. Furthermore, it follows that the magnitude of short-term advective changes is large where the horizontal gradients of a given environmental property are large (*eg*, at current convergences and in coastal areas). The relative magnitude and duration of short-term changes of sea surface temperature (*Fig 62*) show that these short-term changes can exceed one quarter of total annual range of the change of sea surface temperatures at any given location and even, in some occasions, they can exceed half the annual range of monthly mean values in a few days.

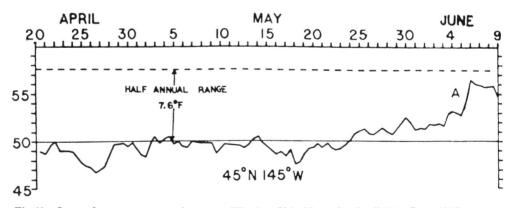

Fig 62   Sea surface temperature changes at Weather Ship *November* April 20 to June 1965.

132

There are profound changes in availability of many pelagic species within short periods at given locations, and these availability changes have been for a long time accredited by fishermen to changes of weather and weather-caused changes in the sea. Although short-term changes in the ocean can be forecast from weather predictions, more research must be conducted on the relation between short-term meteorological and oceanographic changes, and the behaviour and availability of fish species affected by such changes. Experiences have also indicated that the behavioural characteristics of a given species to short-term changes may vary in different geographic areas.

8.3 Climatic changes

Summary

Fluctuations of various climatic parameters occur in the atmosphere, such as seasonal mean position of high and low pressure centres, air temperature, and winds. These climatic changes are associated with changes in surface temperature and surface currents in the ocean.

The climatic changes in the surface layers can vary over relatively short distances in coastal areas. In the open ocean the climatic changes are regional and can be contrary on different sides of the oceans. The interannual variations in environmental parameters are considerably larger than relatively gentle longer-term climatic changes.

It is difficult to obtain reliable climatological data which are consistent in horizontal scale and not contaminated with local anomalies.

Cushing and Dickson (1976) have summarized some known aspects of climatic changes, the response of the ocean on these changes, and some observed changes in distribution of some marine animals (including fish) due to these changes.

Undoubtedly some changes in distribution and abundance of some populations have been caused by minor climatic changes. On the other hand, it is difficult to account for many observed changes as being caused by climate – because due to short periods of observations they might be coincidental with climatic changes, although the reasons for these changes are manifold, as many factors affect the changes of abundance and distribution of fish.

Studies of long-term, and consequently large-scale environmental changes in the oceans suffer from the lack of adequate environmental data bases. An indication of long-term trends in oceanic conditions is afforded by annual mean air temperatures at coastal sites. As an example of coastal air temperature changes, the data at Sitka (*Fig 63*) extend back (with gaps) as far as 1828 and reflect: three to five year temperature cycles, pronounced warm periods (1828, 1869, 1885, 1915, 1926, 1940-41), recent cold periods (1955, 1971), and extended cooling trends (1828-1850, 1926-1955). The longest records of oceanic conditions are those of surface temperatures from 'ships-of-opportunity'. The earliest quasi-periodic, oceanic, sea surface temperatures in the Pacific are available from

133

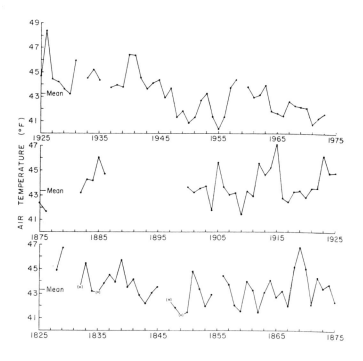

Fig 63   Annual mean air temperatures at Sitka (new
Arkhangel'sk) from 1828 to 1974 indicating short- and long-
term environmental fluctuations.

the trans-pacific voyages of the *China Steamer* from San Francisco to Yokohama for
1871-75 (from Favorite *et al*, 1977). Comparisons of data off San Francisco for the same
months in different years reflect differences of 1-2°C, but the data series is too short to
allow assessments of long-term changes. The accumulation of temperature data from
various vessels has permitted the compilation of numerous atlases of mean conditions,
the most recent and best being that of Robinson and Bauer (1976) showing mean
monthly values at several depth levels, as well as other data analyses. However, in most
instances, the data base for temporal and spatial comparative purposes over large oceanic
areas is limited to surface temperatures, extends back essentially to 1930, and has
reasonable continuity only when averaged in 5° × 5° quadrants.

Within the last several decades there are three generally recognized periods of
extreme anomalous sea surface temperature conditions in the northern Pacific Ocean:
    (*1*) cold conditions in the vicinity of Japan 1934-36,
    (*2*) warm conditions along the west coast of the United States in 1957-58, and
    (*3*) cold conditions in the eastern Bering Sea and off the California coast in 1970-71.
    When all sea surface temperature data are compiled into annual means by 5° × 5°
quadrants (*Fig 64*) these, as well as other, cool and warm periods are evident; certainly

134

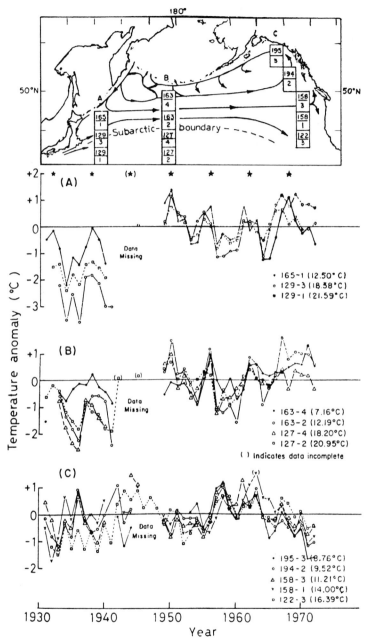

Fig 64 Deviations from annual mean sea surface temperatures in the numbered 5° × 5° quadrangles, 1930-1972 indicating the long-term variability, and the transpacific continuities and discontinuities of specific events (from Favorite and Ingraham, 1976).

135

the cold conditions off Japan in the mid-1930s are striking. The cool conditions in the 1930s were a transpacific phenomena with cold conditions occurring at both sides of the ocean from 1932-34.

If one considers only the years 1953-60 and ascertains the years of maximum positive anomalies in individual $5° \times 5°$ quadrants (*Fig 65*), there is an apparent, orderly eastward shift in maxima from 1955 to 1958 that results in anomalous warming all along the west coast from California to the Gulf of Alaska. Thus, it is obvious that there are marked long-term changes in the environment of the northeastern Pacific Ocean, and these are transpacific phenomena whose causes are not known at this time. There are speculations that the warming is caused by an increase in transport in the warm Kuroshio Current and that cooling may be due to an increase in transport in the cool Oyashio Current. Since the former is a northward flow and the latter a southward flow, these conditions would reflect the more dominant influence of the former on conditions along the west coast of North America. Similarly, the cause for the recent cooling (1970-71) along the west coast of North America is not known.

Climatic changes in the North Atlantic have been described by Rodewald (1960), Cushing and Dickson (1976), and by others. Analysis and comparison of these reported climatic changes in different northern hemisphere oceans reveals that these changes are considerably different in different parts of the oceans. Furthermore, the coherence and nature of the changes can vary over relatively short distances. This is illustrated in *Fig 66* with a number of sea level stations from the Gulf of Alaska. Sea level changes depict changes in local surface pressure, wind direction and speed, and are related also to

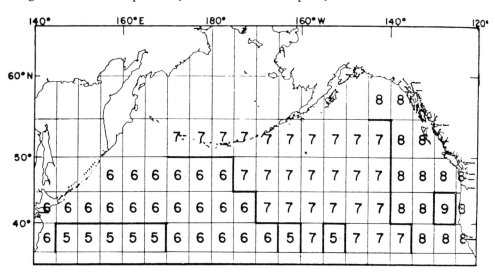

Fig 65   Map showing years from 1953 to 1960 (numbers indicate last digit, 7 =1957) in which maximum positive anomalies (1948-67 mean) from annual mean sea surface temperature occurred in individual $5° \times 5°$ quadrants, indicating the transpacific advection of warm conditions from 1955 to 1958 (from Favorite and McLain, 1973).

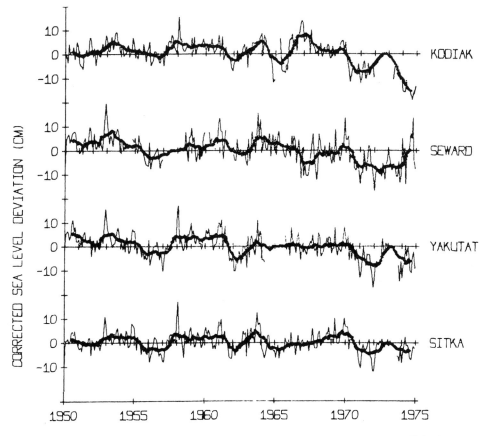

Fig 66 Deviations in corrected sea level (cm) at Kodiak, Seward, Yakatat, and Sitka from monthly mean (1950-74) values (bold line indicates 12-month running mean) showing both coherence and lack of coherence between stations, and marked short-term (monthly) and long-term changes (Ingraham *et al* 1976) (from Favorite *et al*, 1977).

currents and temperatures. The long-term and year-to-year anomalies do not mean that the given environmental condition has been anomalous throughout the year. The anomalies are usually caused by anomalies in one season as *Fig 67* demonstrates. In this example the wind speed anomalies in the twelve-month running mean have been caused mainly by the wind speed anomalies during the winter season.

The effects of the anomalies and climatic changes on the fish biota are difficult to interpret as will be briefly shown in the next chapter. It is often more meaningful to compare monthly means of a given environmental parameter between different years, as demonstrated with storm tracks on *Figs 68* and *69*. The great differences in the storm tracks in March of 1965 and 1967 influenced the fish availability and fishability and thus monthly landings in Icelandic-Norwegian Sea area more than any possible climatic anomaly *eg* in sea surface temperature.

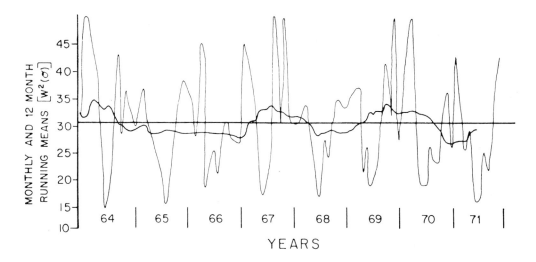

Fig 67 Monthly mean standard deviation of surface wind speed square and 12-month running means off the Washington-Oregon coast from 1964 to 1971, indicating the variations in winter wind speeds.

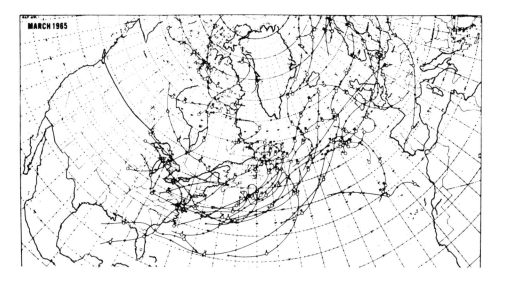

Fig 68 Storm tracks at sea level in North Atlantic, March 1965.

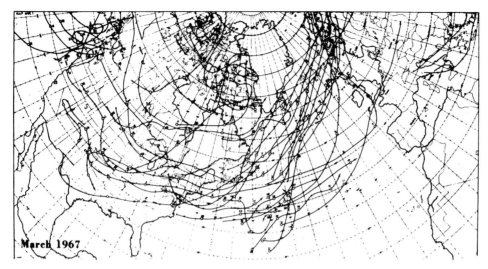

Fig 69   Storm tracks at sea level in North Atlantic, March 1967.

Finally, we emphasize the difficulties of obtaining reliable climatological data for comparison with fisheries data. The climatological data from coastal stations can vary considerably over short distances (*Fig 66*) and is usually not representative of changes which have taken place further offshore. Bell and Pruter (1958) pointed out that coastal air temperature anomalies do not always reflect the sea surface temperature anomalies, and that it is not possible to interpret sea surface temperature anomalies as anomalies of subsurface environment where the fish live. Furthermore, the location of the observation station has not always been the same in all series and local topographic influences might be considerable. The climatological observation series are often broken and incomplete and comparison from isolated years are not much use in determining trends, as temperatures and other climatological parameters in individual years may vary greatly in any period.

9

Effects of ocean variability on marine biota

Summary

The short-term response of fish to short-term variability of the environment is manifested most often in vertical (depth) movement. The horizontal response is usually slow and only in response to long-term changes. Greatest horizontal effects are advective. Greatest effects of ocean variability on fish is usually during the spawning season. The most pronounced short-term environmental variability is in surface weather which affects the fishability as well as availability of fish.

A number of catch prediction schemes have been established for different species and areas based mainly on empirical relations between water temperatures at given stations or sections or wind data and catch in nearby regions.

Many fish-environment relations are masked by economic and fishing practice changes. Before satisfactory relationships can be established between climatological changes and the productivity of fish stocks a great deal more must be known regarding both the environment and the fish themselves.

A need exists also to evaluate the hierarchy of fish responses in cases where several environmental and physiological factors affect the fish simultaneously.

There is a broad spectrum of time and space scales, from seconds to decades and centimetres to thousands of kilometres associated with environmental changes that may be expected to influence the distributions, migrations, and aggregations of fishery resources. The research and verification of the effects of environment variability on the abundance and temporal and spatial behaviour of fish has been scant in most ocean areas and some of the examples in this section have been derived from rather indirect evidence.

The catches of different species vary widely in time and space, and some of this variability can be associated with environmental causes as shown schematically in *Figs 35* and *36*. Neither the causes for migrations nor the actual migration paths of fish are always clearly understood. Tagging studies indicate that even some demersal species which were believed to be rather stationary are undertaking extensive migrations, such as English sole in the North Pacific. Some of these migrations may be triggered by the availability of proper food in the lower levels of the food web (*eg* zooplankton) which is relatively sensitive to environmental changes.

One of the more practical aims in studies of environment-resource interaction is to predict the abundance and availability of fish from analyses of environmental conditions because the environmental variables are easier to measure than the fishery resource itself. The effects of the environment on the behaviour and availability of fish has been described throughout the previous chapters. Here we will examine briefly whether various aspects of fish behaviour can respond to the variability of the ocean in various space and time scales corresponding to the environmental variability described in previous chapters.

Since any change in the environment has time and space dimensions, we must ask whether the speed of the response of the fish can be in phase with the environmental change. As the vertical gradients of environmental properties are sharp and as most pelagic and semipelagic fish undertake vertical migrations (*eg* diurnal migrations), they can respond to any environmental change by adjustment of depth. Obviously the fish has the option to move horizontally in response to an environmental anomaly. Depending on the gradient present in any given location, a 1°C anomaly can mean the north-south displacement of the isotherms of 10 to 200 nautical miles. There is no proof that fish populations have reacted with horizontal migrations to short-term environmental anomalies, except in the course of their seasonal and spawning migrations. Thus, the horizontal response to environmental variability might be considered to be slow.

The environmental anomalies can, however, have an effect on spawning and its success, as discussed in Chapter 3 §3.1. Other 'slow' effects of environmental changes on fish populations are via the effects on growth and on predation through changes in predator-prey distribution. Thus, the synoptic variability can affect mainly the distribution and behaviour of fish vertically (within depth). Seasonal and longer term variability can affect the horizontal distribution and abundance of fish with some time lag. These long-term effects might be used for various fisheries forecasts as will be briefly summarized at the end of this chapter.

The synoptic and medium-range variability of surface weather can affect fishing and, since the fishery provides the main data source for evaluation of changes, can thus mask effects in the ecosystem. For example, considering the differences in storm tracks between March 1965 and March 1967 (*Figs 68* and *69*), it is probable that trawlers from Europe carried out more fishing off Lofoten or in the Barents Sea in March 1965 than on the Grand Banks.

It could be assumed that one of the main factors affecting the year-to-year variations of the availability of fish on any given ground might be the prevailing current which is affected by the year-to-year anomalies of winds. If so, the methods suggested for the prediction of the brood-strength fluctuations (see Carruthers *et al*, 1951), could also be used to predict the availability of adult fish. Craig (1958) has shown that the availability of herring four years old and older in the Buchan pre-spawning fishery is

141

related to wind and temperature during the catching season. Obviously, the temperature can be greatly affected by advection and, thus, the whole fishery might be directly related to the currents which in turn are related to the winds. Craig (1958) devised a formula based on the advection and on the temperature ranges to predict the catches of herring in the Buchan fishing grounds. The main basis for such a prediction was the analysis of wind anomalies.

Rodewald (1960) analyzed the fluctuations in landings and in the availability of different commercially important fish stocks in relation to the anomalies of wind in the Barents Sea, Labrador area, and Icelandic waters. His results indicate that anomalous water transport, caused by large-scale variations of the atmospheric pressure and winds, determines the availability of a number of commercially important fish species, especially in the northern waters where they are distributed close to the boundary of their normal range. By predicting the pressure and wind anomalies and the resulting anomalous currents, one would be able to predict the availability of fish on these fishing grounds.

In a simple way the weather conditions during a particular fishing season for a given species can be compared with the average climatic conditions during that season over many years, and on this basis the weather can be classified as favourable or unfavourable from a fisheries point of view. This approach has been utilized by Rodewald (1960). One such example is shown as *Fig 70* where the favourable and unfavourable weather conditions for herring in the North Atlantic are shown by comparing the average distribution of atmospheric pressure during the month of August in the year 1952 (cool and stormy – German catch 282 000t) and 1953 (warmer and calmer – German catch 343 000t).

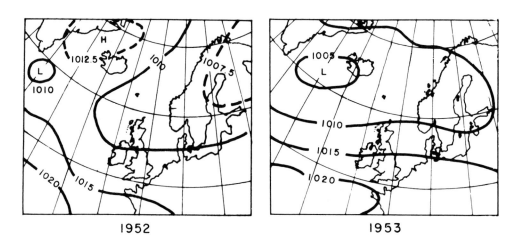

Fig 70　Mean distribution of atmospheric pressure over the North Atlantic in August 1952, which was a bad bank herring fishing year, and in August, 1953, which was a good year (after Seilkopf).

142

A more detailed approach to the problem of the relation between meteorological factors and fish distribution has been made statistically by correlating the individual catches of herring by German luggers in the North Sea with the wind conditions observed on shipboard during the catching time. The partial results obtained so far seem to confirm the conception of many fishermen that there exists a relation between the wind direction and the catches obtained (Walden and Schubert, 1965). The monthly or seasonal anomaly of the barometric pressure is sometimes a suitable parameter for the interpretation of causes of the fluctuations in the yield of fishing. The pressure anomalies act upon the surface currents and their boundaries through surface wind anomalies, and thereby influence the advection of cold or warm waters in an anomalous way. According to Rodewald (1960), in 1954 the water temperature in the Barents Sea was relatively high because of the 'anomaly winds', which forced more warm water into the area and prevented the cold water from entering from the northeast. The fishing in the Barents Sea was good. From 1956 to 1958 a distinct cooling of the Barents Sea was brought about by 'anomaly winds' from the east preventing the warm Atlantic water from entering and forcing the cold arctic water into the area from the northeast (see *Fig 71* with the years 1954 and 1957 as examples). The fish, especially cod, haddock, and saithe (*Pollachius virens*) were forced to the west, and the fishing in the Barents Sea was very poor; indeed, the fishing by some European countries almost ceased in the area. In 1959 the wind conditions were similar to those of 1954 and the fishing was successful again.

It is both simpler and also in many cases more appropriate to correlate single environmental factors to certain selected aspects of the fish behaviour and availability. It

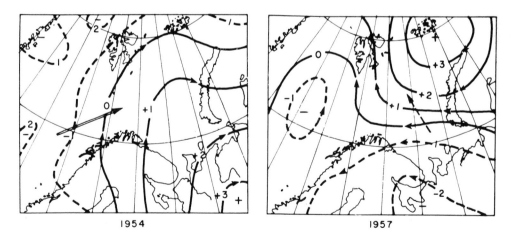

1954   1957

Fig 71 Favourable and unfavourable weather conditions for the cod fishing in Barents Sea, presented as anomalies of the atmospheric pressure for the years 1954 and 1957. Arrows indicate the anomalous transport of warm and cold water for respective years (after Rodewald, 1960).

would not be practical to analyse all the meteorologic-hydrographical factors and all their marine-environmental effects as one problem-complex. One of the most easily measured and observed environmental factors, as we have seen, is the sea temperature, which is of course a function of several meteorological factors: the amount of radiation penetrating the sea surface, difference between sea and air temperature, the advection of heat by currents, winds, *etc*. There are long-term temperature fluctuations in the sea which manifest themselves in the north- or southward migrations of economically important stocks; the cod in Greenland waters is a classical example. Besides this may be mentioned the northward movement of sardines in the English Channel and southern North Sea and the more frequent occurrence of tunas in the North Sea from 1951 onwards, apparently related to a slight increase in temperature.

During the last fifteen years several fisheries forecasting schemes have been proposed which are based on the response of fish to environmental anomalies. We summarize a few of these schemes below to illustrate the possible response of the resources to long-term environmental variations and anomalies.

Uda (1974) concluded that a general relationship exists between the environmental conditions, particularly current patterns, and the availability of commercial fish species in Japanese waters. He proposed a long-term forecast for fisheries, based mainly on prediction of changes in current patterns (Kuroshio, Oyashio, and Tsushima Currents).

Three fisheries forecasting schemes have been proposed for Okhotsk Sea herring. Chernyavskiy (1970) used water temperature in the surface layer to indicate areas of best daily yields. Yields of more than 200kg/net were expected in the areas where temperature was 7·5 to 10·5°C, whereas no herring catch was expected where temperature was < 4°C or > 13°C. Tyurnin (1973) used the water temperature and ice cover during the spawning season as indicators for planning of herring search operations. Elkin (1973) found that the peak spawning and large herring shoal formation was water temperature dependent and proposed a scheme to predict about three months in advance the period of large shoal formation during which shoals were slow moving and easily caught by purse seines.

Four prediction schemes based on environmental anomalies and fish availability relations have been proposed for the Barents Sea. Konstantinov and Svetlov (1974) predicted the catch of bottom fish by trawling two to three months in advance, using water temperatures in 150 to 200 metres depth along a standard observational section. Konstantinov (1973) found that the temperature conditions (and anomalies) in the southern Barents Sea affected the migration of cod to Spitsbergen area and proposed a prediction method for distribution of cod in West Spitsbergen three to six months in advance of the start of trawl fishery there using water temperatures along the standard section in the Barents Sea. A similar approach for prediction of catch of cod and haddock in the Barents Sea, based on temperatures in the same standard section, was proposed by Mukhin and Ponomarenko (1968). The studies of distribution of demersal fish in the Barents Sea by Østvedt (1971) indicate that there is a more easterly distribution of

144

demersal fish in the Barents Sea during 'warm' years. This knowledge was utilized by Ponomarenko (1968) and Konstantinov (1967) to predict the catches in various sub-areas of the Barents Sea some months in advance.

The effect of the water temperature anomaly during spawning, and its relation to year-class strength has been utilized in several prediction schemes. Buys (1959) proposed to predict the size of pilchard catch one year in advance off South African coast using surface water temperature in defined areas as predictor parameter. Antony Raja (1974) found that he could predict the prospect of commercial fishery for the juveniles of Indian oil sardine about two months ahead using mean daily rainfall in June to August, rainfall being taken as 'symptomatic of general climatic conditions that would control spawning'.

Based on the study of long-range catch statistics and water temperature at fixed stations, Sutcliffe *et al* (1977) showed that it might be possible to predict catch of various species (cod, haddock, yellowtail flounder, and menhaden) along the US Atlantic coast with different time lags (one to seven years). Flowers and Saila (1972) believed that the prediction of lobster catch along the US Atlantic coast is possible by taking into account mean bottom temperatures of preceding winter months and mean annual temperatures six to eight years previous to the year for which the prediction is made.

Finally, some words of caution about the use of relations between climatic changes and changes in fish stocks: Bell and Pruter (1958) re-examined a number of reported climatic temperature-fish productivity relationships. They found that effects of other variables, such as economic conditions, changes in fishing practices, and the extent of the removals by man, overshadowed the climatic change-fishstock change relationships.

The possibility of coincidental relationships must be considered by exhaustive tests of the representatitiveness and adequacy of both the environmental data and the fishery data, particularly that of catch per unit of effort. Where the fishery data are defective they should not be used to judge the effects of either the environment or man upon the stocks. Methodology of relating the variables, such as the time-lagging of data and the use of correlation analysis, must conform with the known facts about the species investigated.

Normally several environmental factors affect the fish simultaneously and questions arise: to which stimulus does the fish respond; which is the strongest signal? There are few experimental results bearing on such questions. As an example, Balchen (1979) proposes a theoretical method for simulation of a multiple response in an ecosystem model – *eg* a resultant migration affected by two simultaneous stimuli, temperature and food availability gradients.

10

Ocean analyses and forecasting and fisheries diagnostic/ prognostic services

Summary
Ocean analyses and forecasts present the distribution of environmental parameters in space and time. These analyses can be used in fisheries as partial guides to (*a*) estimation of the probable abundance and distribution of fish, (*b*) prognostication of anomalies in migration, aggregation, dispersal, and other behaviour, and (*c*) estimation of anomalies of spawning conditions and spawning success.

Most of the ocean analyses for fisheries applications are done by meteorological services. The contributions by fisheries research to oceanographic research have declined in the last decade. Fisheries research is, however, making extensive use of environmental monitoring from coastal stations and along fixed sections.

Sea surface temperature analyses and wave predictions are the most common oceanographic analyses prepared by these services. The thermal structure is conventionally measured with BTs, but could also be predicted numerically if application of such data can be demonstrated to be cost-effective. Surface currents and waves can also be estimated with relatively simple empirical approaches. However, advanced numerical computerized models (such as HN models) are available for accurate analysis and prediction of currents, and special wave models can be used for wave prediction.

Application of remote sensing of sea surface parameters from satellites have been limited to monitoring sea ice but other applications based on surface parameters may be anticipated. Additional environmental predictions for fisheries applications are possible but could be routinely issued only if they (*a*) result in economic gains, and (*b*) contribute materially to resource conservation and management.

The basic objective for fisheries environmental services would be to assist fishermen in planning fishing operations and in searching for harvestable fish concentrations. These services would be of potential economic value by reducing search time and operating costs for fishing boats and their crews.

A presupposition of such services, however, is that meaningful relations between the harvestable fish concentrations and some easily observable environmental parameters are known. It would also be desirable to understand the effects of these changes on the reproductive success of fishes and on fluctuations in the number of recruits. Such knowledge is not yet available for some fisheries – and more research is needed on migrations, spawning, wintering places, distribution in the water column, *etc* in response to changes in the ocean – but for others meaningful relationships between abundance and availability and oceanic parameters have been found. These may be used by environmental services to raise the efficiency of fishing operations and to improve long-term management of fisheries.

Some detailed objectives and use of fisheries environmental services for practical fishing are:

——prediction of distribution and abundance of a given stock as determined by optimum and/or limiting environmental conditions,

——prediction of dispersal, aggregation, and vertical migrations as determined by the synoptic conditions,

——prediction of direction and speed of migration, depth of occurrence of fish shoals, and estimation of the date of arrival of species to their spawning grounds,

——forecasting weather and sea conditions as affecting fishing operations.

In addition, effective fisheries management requires:

——estimating natural fluctuations of fish stock as determined and influenced by environmental changes.

The monitoring and prediction of the environment, specially the weather, is done by national meteorological services. A few ocean environmental parameters such as surface temperature and waves are analysed and predicted routinely by such services. However, fisheries research agencies must still be engaged in ocean environmental monitoring and analyses to collect systematically ocean environmental data which are pertinent to fisheries problems.

A detailed review of oceanographic analysis and forecasting methods was given by Laevastu and Hela (1970). Here we review only briefly these methods and the necessary monitoring data. Of the oceanographic observations required for ocean analysis only sea surface temperature, wave height, and surface wind are obtained routinely. Most water column measurements are made on an *ad hoc* basis as part of specific cruises and these data are obtained over relatively wide geographic regions. Seldom is the number of observations reaching an environmental service centre within a short time period sufficient to prepare an analysis which represents the actual synoptic distribution of oceanographic parameters at depth.

For fisheries purposes there is often a need to establish coastal monitoring sites to discover any profound anomalies. As environmental monitoring is a large task, to be worthwhile the subjects to be monitored must be selected with careful consideration of space and time scales and their applications. In synoptic monitoring we are often mainly concerned with sudden, more profound events (such as storms) which trigger changes in the marine ecosystem. In medium-range (seasonal) monitoring, the concern is mainly with environmental anomalies, and in long-range monitoring, the trends of changes are important.

The main source of the synoptic sea surface temperature (SST) data is the marine weather reports from voluntarily observing and reporting vessels.

Two differing basic methods for measuring the SST on these vessels are in use: the bucket method and the 'condenser' (cooler) intake thermometer method. Numerous studies have been made on the accuracy, sources of errors, and resulting differences of these two methods. In general the mean error is about $\pm 0.7°C$ but the intake temperatures are about $0.2°C$ warmer than the bucket temperatures. Besides the different reading and handling errors in observations of the two types, their rather large inaccuracies are often due to the fact that the thermometers for measuring the SST are not checked and calibrated by the port meteorological officers.

About 1 200 SST reports are available every 12 hours from the northern hemisphere. This data density is too low for reliable synoptic analysis and thus the period of 12 hours is too short. As the SST changes are not too sudden in most areas, it is feasible to use data from a period of three and a half days in an analysis, provided that the analysis scheme allows some indirect weighing of the data by their age. *Fig 72* shows the SST data density during a given period of three and a half days in March 1967. This chart indicates that the data density is reasonable between 25°N and 60°N approximately.

The synoptic BT and XBT (expendable bathythermograph) reports necessary for subsurface temperature analyses originate from research vessels, ocean station vessels, naval vessels, and from some fisheries and merchant vessels. Some BT observations are lately exchanged internationally via IGOSS system. The accuracy of data obtained by the 'mechanical' BT is limited for a number of reasons, whereas the XBT method, in addition to achieving greater depth, has few limitations when employed as a data gathering instrument.

The determination of the mixed layer depth (MLD) from a single BT observation is always subject to error because of the short-term vertical fluctuations of the thermal structure resulting from the convergence/divergence, internal seiches and tides, winds, and other factors. It should be noted that the average short-term fluctuation of the MLD is about 5m (16ft); however, fluctuations of 15m (50ft) occur in some areas and seasons.

Synoptic oceanographic data can be obtained also from some coastal stations and from research vessels. Recently some of these data have been routinely transmitted by radio messages, but much remains to be done to make effective use of other observations.

Truly synoptic oceanographic observations are scarce, consequently the variability of the ocean conditions must be derived using knowledge of their relation to meteorological driving forces. Furthermore, many resource-environment problems require the direct use of specific surface meteorological data, such as winds. Modern meteorological

SEA SURFACE TEMPERATURE THREE AND ONE HALF DAY COVERAGE MRO TEN 80330367   7398 REPORTS

Fig 72 The density of synoptic sea surface temperature reports during a period of three and a half days, ending at 00Z, 13th March, 1967. The numbers on the chart indicate the number of reports in each area of the character. 9 means 9 or more reports.

149

analyses and forecasts are usually accomplished using computerized models. The earlier, simpler barotropic/baroclinic models are now replaced with primitive equation (PE) models, using hemispheric grid nets with grid size of about 380km. Such models not only speed the analysis process and reduce manpower, but carry out the task objectively with prescribed procedures, eliminating subjective interpretations which may vary from person to person.

Over open ocean areas the hemispheric, 'standard mesh' models are sufficient for fisheries oceanography purposes. In these areas the surface meteorological features are large in scale and the available synoptic data from voluntarily observing and reporting vessels are sparse. In coastal areas, where most fishing is conducted, the meteorological conditions are more variable than further offshore, and are influenced by local factors, such as mountainous coasts and intensive sea-air interaction over coastal waters. Thus, the dimensions of the significant environmental features, such as wind and temperature fields, are considerably smaller in coastal areas and the coarse mesh hemispheric models are not adequate to analyse and predict this environment. Consequently, small-mesh numerical models must be used for meteorological and oceanographic analyses and forecasts in coastal areas.

Although excellent small-mesh meteorological forecasting programmes are in existence, they are seldom used; one reason being the absence of enough demand for these analysis/forecasts from the fishery community. The small mesh analysis/forecasting models, with grid size from 50 to 125km, are usually baroclinic models, utilizing among other factors, quantitative knowledge on sea-air interactions. The boundary conditions, and at times the initial guess fields, are derived from hemispheric models. In a forecasting mode these models are useful and more accurate than hemispheric PE models for periods up to about 48 hours. Thereafter, the prescribed boundary conditions (ie, the atmospheric systems which move into the area from the boundaries) dominate the prediction.

The sea surface temperature (SST) is one of the most easily measured environmental properties in the sea. It is the only truly oceanographic element which is satisfactorily observed and reported on a synoptic schedule. The information on the SST is useful *per se* for solving a variety of problems and, in addition, the SST is indicative of other conditions and processes in the sea. In developing truly synoptic oceanography, the importance of the knowledge of the SST distribution is comparable to the importance of the knowledge of pressure distribution in the atmosphere.

The methods for analysis of sea surface temperature (SST) use SST reports from meteorological messages and, in special cases, observations by fishing vessels. The satellite infra-red observations, as well as heat exchange computations, have potential but are not yet used for this analysis due to their low accuracy. Due to low density of observations and since SST changes relatively slowly, three to seven days composite observations can be used in each analysis. Information in surrounding areas can be obtained by interpolated climatology. The grid size in any oceanographic analysis should normally not exceed 50km as oceanographic features and processes have slightly smaller space scales than corresponding atmospheric features. An example of computerized synoptic SST analyses is given in *Fig 73*. The SST analysis is usually subtracted from time interpolated SST climatology to obtain anomalies, which have more direct use in fisheries ecology problems than absolute SST values.

At times the temperature itself may not be the directly affecting factor we are looking for, but it can be used since it may indicate other changes and conditions in the sea. Examples of indirect use of SST are estimation of upwelling intensities and the computation of current (and surface water type) boundaries.

Various methods for the analysis of the subsurface thermal structure and their scientific background are described by Laevastu and Hubert (1965). The steps in a

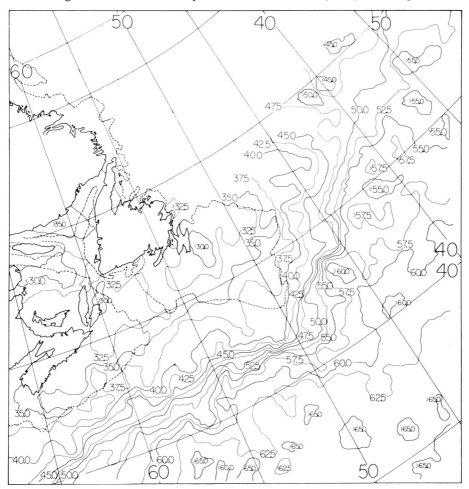

Fig 73 The synoptic sea surface temperature analysis of the Grand Banks area at 00Z, 17th March, 1967.

151

manual and computerized analysis of the subsurface thermal structure are essentially identical. However, due to the labour involved, manual methods do not allow for the detailed consideration as do those with the computer. Consequently the manual methods are vastly simplified versions of the numerical ones.

The analyses of mixed layer depth (MLD, depth of the thermocline) and surface currents can be computed, using mainly surface wind data from meteorological analyses, often augmented by auxiliary data from ocean climatology. The methods for these computations are relatively complex and their description can be found in Laevastu and Hela (1970). The few directly synoptic observations on MLD [bathythermographs (BTs), expendable bathythermographs (XBTs), or temperature-salinity-depth (TSD) records] and surface currents are usually used to refine analysis/prediction methods, rather than used directly in synoptic analysis.

It should be noted that the thermocline can fluctuate up and down considerably in 24 hours. This fluctuation is predicted partly on a physical and partly on a statistical-empirical basis. It should also be noted that a single BT cast is not an absolute measure of MLD and sub-surface thermal structure as there is no means at present to determine at which stage of the fluctuation of the thermocline the BT cast was taken.

The advanced, multilayer Hydrodynamical Numerical (HN) models compute the currents (including tidal currents) and the thermohaline structure at each grid point in each time step. They require the use of large computers, however, and are expensive to use. As these models can depict in detail the distribution and dispersal of various components of the water masses, they could be used in special research applications to describe distribution and dispersal of fish eggs and larvae or pollutants and in more accurate research tasks. Furthermore, the HN models can provide environmental input to ecosystem models, especially where tidal currents, mixing by tides, and net tidal drifts are of concern.

The wave heights and periods are observed at synoptic hours and reported with the meteorological data by the voluntarily observing and reporting vessels. These reports are used for the construction of synoptic wave charts. Numerous methods and rather voluminous literature exist for wave forecasting. The basic theories and description of the wave problems are found, for example, in Sverdrup and Munk (1947) as well as in the textbooks on oceanography. Wave forecasting over an ocean area can either be made manually or by using computers. For prognostic work over large areas manual methods of forecasting for each given point are too complex and time consuming, and, therefore, simplifications are necessary.

Simple 'singular' empirical methods for wave height estimation which relate wave parameters to present and near past surface wind parameters have been used for some time in the past. They are rapid and useful for estimating the height of a fully risen sea in a few given locations on a weather map. They serve at times for auxiliary computation in other forecasts, such as, the determination of forced mixing and in the estimation of the thermocline depth.

Lumb (1963) developed a simple method for forecasting the maximum wave height in the North Atlantic based on the analyses of wind and wave data from weather ships $\mathcal{J}$ (52°30′N, 20°00′W) and I (59°00′N, 19°00′W). The three straight-line relationships between the maximum wave height and the wind speed are given in *Fig 74*. The formulas for those lines are as follows:

Line I	$H_{max} = 1·16\,V - 3·3$	(10)
Line II	$H_{max} = 0·90\,V - 1·5$	(11)
Line III	$H_{max} = 1·20\,V - 8·7$	(12)

The mean wind speed V (in knots) was obtained by assuming a steady change between the 6-hourly reports:

$$V = \frac{V_1 + 2V_2 + V_3}{4} \qquad (13)$$

It should be noted that the line I in *Fig 74* is valid from September to April if the cold air prevails over the warmer water, and the line II for the same period if the air is warmer than the sea surface. The line III is valid from May to August. Lumb's

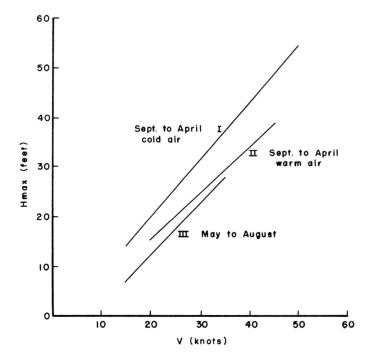

Fig 74 Relations between the maximum wave height and the wind speed (after Lumb, 1963).

153

empirical relations are not applicable for winds weaker than 20 knots, since with light winds the swell often dominates over the sea and the simple relationship between H_{max} and the average wind speed, presented by the straight line, then breaks down. For example, the use of the line II could lead to large errors if used for the hours after the passage of a warm front or a warm occlusion when strong winds are replaced by weak mild air flow.

According to Lumb, the simple relation between H_{max} and the average speed gives strong support to the opinion, often expressed by shipmasters, that if the surface wind direction and its speed are correctly forecast, long experience enables them to judge quite accurately what the sea conditions will be, except in the areas where the swell will dominate.

The causes of surface current can be manifold. The speed and direction of the current, at a given point, time, and depth (*eg*, at a depth of 3m), W_{ptz}, can be represented as the resultant of the following components:

$$W_{ptz}=G_c+W_c+W_m+P_i+P_t+A_c+I \tag{14}$$

where:

$\quad G_c$ – permanent flow (gradient current and/or 'characteristic current'),

$\quad W_c$ – wind current (lately often erroneously called the 'Ekman current'),

$\quad W_m$ – mass transport velocity by waves,

$\quad P_i$ – periodic portion of inertia current,

$\quad P_t$ – periodic portion of tidal current,

$\quad A_c$ – current caused by changes of atmospheric pressure and sea level,

$\quad I$ – the velocity and directional component, caused by influencing factors, such as changing depth of water, Coriolis effect, and coast and current boundaries.

Considering the above formula, it becomes apparent that surface currents should be (and are) computed with complex computer programme. At times it is useful to estimate surface current from surface wind data alone. Witting (1909) formula is useful for this purpose.

$$W_w=k\sqrt{Wg} \tag{15}$$

where W_w is current speed in cm/sec, Wg is average geostrophic wind speed in last 36 hours in m/sec and $k=3\cdot8$. The surface current is assumed to flow in the same direction as geostrophic wind or *ca* 15° to the right of true surface wind. (If true surface wind speed is used, the factor $k=4\cdot4$.)

Tidal currents usually prevail in water over the continental shelf. The computation of the tidal currents with reasonable accuracy is possible using numerical hydrodynamical equations and large high-speed computers (Hansen, 1966). An example of these computations is shown in *Fig 75*. These models utilize tidal harmonics at the open boundary, winds, and actual depths. The models are relatively complex and their description falls outside the scope of this book.

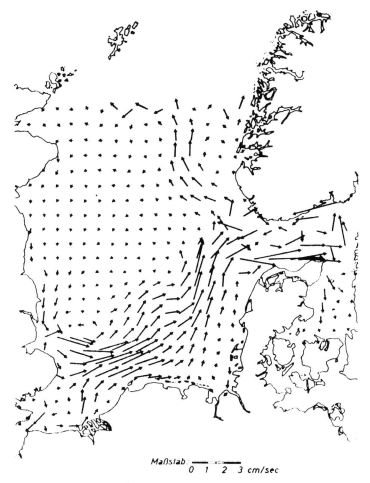

Maßstab: 0 1 2 3 cm/sec

Fig 75 An example of the computed tidal currents. (North Sea, M_2-component). (Hansen, 1966).

Accuracy in oceanographic forecasts is, obviously, greatly dependent on the accuracy of the input of observations and of meteorological forecasts. The standard deviation of each element varies considerably in space and time, as well as with the scale of the analysis. As approximate general limits of the accuracy, the following numbers are presented:

Sea surface temperature	0·3 to 1·8°C
Temperature below thermocline	0·3 to 1·3°C
Mixed layer depth	7 to 20 metres
Wave height	0·4 to 1·5 metres
Current speed	0·1 to 0·4 knots
Current direction	10° to 50°

155

The use of some remote sensing techniques from aeroplanes in fisheries research and commercial fisheries lead some scientists to speculate about their potential use from spacecraft (Stevenson and Pastula, 1971). Although there has been considerable talk about satellite application in oceanography and fisheries, practical results have been few to date. At present ice fields and sediment-laden river plumes discharged into the coastal waters can be delineated from satellite imagery. The observations of sea surface temperatures from satellites are as yet little used due to low accuracy and data contamination by cloud cover and humidity of the air. Remote sensors cannot penetrate through the surface to any significant depth for subsurface observations. Recent observations of surface colour from satellites suggest that this tool may become useful in study of ocean fronts, river plumes, and mixing processes.

It can be concluded that satellites will not replace traditional approaches in fisheries and oceanographic research, and there is little hope of being able to monitor and assess fish stocks through space observations. The greatest potential for use of remote sensors from spacecraft in the near future is in the sensing of sea surface properties, such as waves, ice, colour, sea surface temperatures, and sea elevation. Use of satellites in marine activities for communication and for tracking and interrogating drifting and/or moored instrumental buoys will continue to increase.

Synoptic oceanographic services for fisheries applications in most cases are connected closely with national meteorological services for economic and organizational reasons. These services prepare analyses and forecasts for shipping, among which storm warning and wave forecasts still are the most prominent products. Among these efforts the Japanese meteorological and fisheries agencies have been providing the greatest variety of services to their fishing fleets. According to Terada (1959), who has described the Japanese fleet operations, the fishing fleets also forward synoptic meteorological and oceanographic observations to the Meteorological Agency. The positions of the individual boats are coded, whereby the code is known only to the corresponding fleet and to a few officers in the Meteorological Agency. The Agency prepares four specific forecasts a day for the salmon fishing area which are both broadcast by radio and sent by facsimile. These forecasts include gale warnings, wind force and direction over a large area, positions and intensity of low pressures, wave forecasts, sea surface temperature forecasts, and long-range (seven-day) meteorological forecasts.

In the early 1970s a relatively great number of sea surface temperature analyses were prepared by different countries on different space and time scales (see Tomczak, 1977). Experience gained by various fisheries environmental services (Østvedt, 1971) has shown that these surface temperature maps are not in all cases sufficient to interpret the behaviour of fish, especially its distribution or aggregation in the area of investigation. Consequently, by the 1980s fewer of these analyses are being continued. It now appears that more systematic use of bathythermograph information to explore the subsurface temperature structure (Laevastu and Johnson, 1971) is needed to complement the information contained in SST charts.

It would be impractical to utilize a great number of hydrographic properties as parameters in synoptic services for fisheries. Therefore, proper application of synoptic oceanography requires knowledge of both the interaction between the environment and the fish and also knowledge of critical environmental parameters for a given fishery operation.

A need for fisheries diagnostic/prognostic services can be considered realistic if it: (a) results in economic gains to the fishermen and the processing industry, and/or (b) contributes materially to conservation and management of the resources. Although the needs for such fisheries services have been pointed out in the past and it has been postulated that gains will result, there has been no real evaluation of the resulting benefits on cost-effectiveness bases. Unfortunately the value of such services cannot be measured by short-term economic returns, but only by successful conservation of fishery resources combined with their optimum utilization over time. Long-term national and international interests dictate the conservation of the fishery resources for the future, but demand also their present optimum utilization. Thus, it can be expected that more use will be made of fish-environment relations in holistic ecosystem models for fisheries management purposes (see Chapter 11).

Due to increasing amounts of various fisheries data and the need to use these data in quasi-real time, in various fisheries management problems, it is necessary to computerize the storage, retrieval, and analysis of environmental, fisheries, and biological data. There-fore, certain standardization (specially in respect to format) of numerical storage and retrieval models is in progress in fisheries centres. The models for objective analysis of randomly spaced data in space and time are an important part of these data handling programmes. Some of these analysis programmes were initially developed for environmental observations but have been adapted for objective analysis of great variety of special data, such as catch statistics and results from exploratory fishing.

Two developments in the data handling field with respect to fish-environment relations seem to be in need of additional attention and acceleration: first, fisheries biological data, including catch, effort, and economic data, must be put into computer accessible form (environmental data is generally available in this form) so that these data, among others, can be incorporated into ecosystem models. This objective has already been accomplished in many fisheries laboratories. Second, various computerized analyses should be carried out on various fisheries data in association with corresponding environmental data to test theories and hypotheses on resource-environment relations, to extend this knowledge, and to create a basis for various fisheries forecasting methods.

11

Numerical ecosystem simulations in fisheries research and management

The marine ecosystem, which is invisible to the human eye, is complex in respect to its species composition as well as in respect to processes occurring within it.

The marine ecosystem is not stable, but considerable fluctuations in abundance and distribution of many species occur. The determination of abundance and fluctuations in abundance and distribution of commercially important species and factors controlling these fluctuations is one of the main tasks of fisheries scientists. On the other hand, the complex processes controlling the abundance of species in the marine ecosystem run a steady course over long time-periods. It is quite remarkable that from about 200 000 eggs, released during spawning time by a female pollock, an average of only 2 (a male and a female) fish survive say to age of four. However, deviations from this remarkably constant process of reducing the numbers of survivors occur, so that a given year-class strength of any given species can be few to few tens of times higher or lower than the average.

The fluctuations in the abundance of species can be caused by numerous factors, such as environmental anomalies and/or factors inherent in the populations themselves (*eg*, predation and cannibalism). Although the populations of some species can decrease and others increase with time, the standing stock of the total biomass of finfish fluctuates relatively little in the course of time; its abundance being determined by the total availability of food, where the zooplankton and benthos productions take a buffering role, determining the so-called 'carrying capacity' of any given region.

Obviously fishing will cause changes in the abundance not only in the target species, but also in other species not subject to fishing. These secondary changes can be caused by changes in predation. Furthermore, some changes in target species can be caused by other factors than fishery also (*eg*, by environmental anomalies).

In order to form a quantitative picture of the changes and interactions in the marine ecosystem, it is possible to assemble available knowledge of such systems into simulation models for study, using large computers. This chapter attempts to give a birdseye view of one such approach to ecosystem simulation and its application in fisheries management problems, emphasizing mainly the effects of environment in the ecosystem and the use of ecosystem simulations in the quantitative study of species-environment interactions.

158

11.1 The role of phyto- and zooplankton and benthos in production of fish

Summary

The production of basic organic matter by phytoplankton is controlled by a number of factors, such as availability of nutrients, light, turbulence, mixed layer depth, *etc*. These factors vary considerably in space and time and make the measurements and computation of basic organic production difficult and uncertain. The pathways of utilization of this production are also variable in space and time.

Zooplankton, benthos, and nekton (including fish) are direct food for fish (which includes also some cannibalism). Zooplankton and benthos may be considered the production buffers of the finfish biomass and determine the carrying capacity of fish of any region on a large scale.

Quantitative data on benthos (especially on epibenthic crustaceans) and on larger zooplankters (especially on euphausiids) is deficient in most areas, mainly due to great spatial and temporal variability. However, their occurrence as food items permits their quantitative simulation, which, however, needs experimental validation.

It is concluded that predation is one of the most important controlling processes in the marine ecosystem, which also controls any possible surplus production.

It has been customary to start the description of marine ecosystem and production in it with phytoplankton. Many marine scientists have emphasized the role of phytoplankton production as a determinant for fish production and as an important factor in fluctuation of stocks. Furthermore, most of the existing ecosystem models start with phytoplankton production and its utilization. In this subchapter we will re-examine critically these concepts.

The term production has been widely used (mostly misused) in connection with the dynamics of organic matter in the marine ecosystem. In a dynamic situation we must always define the producer and quantify production as a rate, *ie* define the unit of time, amount, and subject (a single organism of given age and size, a group of organisms in defined three-dimensional space, *etc*). The most common use of the term production is in the production of organic matter by plants, mainly phytoplankton in the ocean ('basic organic production' expressed in terms of mg carbon fixed per unit volume or surface area in unit time).

Organic production, in more general terms, then depends first on the type of producers and their production capacity. Thus, in order to estimate production, we must know: the type (species) and the quantity present at the initial time or at any given time step, *ie* we must know the standing stock or standing crop of all the producing species. The standing stock is in itself a dynamic property which varies in space and time and depends on its production (thus, a nonlinear second order interaction exists). Furthermore, the producers are being eaten and die from other causes, creating secondary nonlinear effects.

159

The production capacity depends (under otherwise identical conditions) often also on the age of the individuals in the population. The production capacity furthermore depends on the following conditions:

——Availability of nutrients from which to produce (*ie* carbon, oxygen, and other constituents). The available quantities of these substances should be measured at given locations and times since their availability is dynamic, depending on transport and regeneration of earlier production.

——Availability of energy which is required for production. In basic organic production light energy is required. The energy available on the surface depends on geographic position and time but is also dependent on variable cloud cover. The availability of light energy at depth depends on the depth and the turbidity of the water. Furthermore, the dynamics of the producers themselves influences the turbidity significantly. Different organisms utilize different wave lengths of light and the energy utilization (or the production capability) of each organism depends also on temperature. Because of these variables the space and time variation of light energy can only be approximated at present.

The data necessary for estimation (computation) of basic organic production are seldom available in desired space and time scales and these estimates are at best gross approximations. In fact, most of the parameters necessary for the computation of basic organic production can only be estimated at any given location and time to the order of magnitude. Consequently these estimates, nor the widely varying scattered measurements of basic organic production, do not warrant a complex formulation of the production process nor do they allow computations that would be terminant for other (following) processes in the marine food chain. Thus, it is self-defeating to base quantitative ecosystem computations at higher trophic levels on the estimation of basic organic production and its utilization. We can thus use organic production measurements and estimates only in large space and time scales for comparison of the productivity of different regions.

In the literature the term 'secondary production' usually refers to the dynamics of the zooplankton biomass, but sometimes includes other filter feeders and even fish. The first step in the utilization of biomass is the process of eating (predation, grazing). The second step is the digestion of the eaten matter which results in the breakdown of some to basic nutrients and organic compounds (non-living), conversion of some into new biomass, and excretion of some of the eaten matter which can be used by other organisms, including bacteria.

The conversion of the digested matter into a new form of biomass results in growth via anabolic processes and the breakdown of organic matter by catabolic processes (together termed metabolism). The utilization of organic matter for energy release is also a process of regeneration of basic nutrients. Thus, from the point of view of dynamics of organic matter, we must consider the processes of predation, growth, and re-utilization.

160

Predation depends on many conditions of the prey and predator as well as general ecosystem conditions at large. Among these determining factors are: suitability (preference), predator/prey size relationships, availability (encounter-density dependent), and avoidance behaviour. Thus, mobility and distribution of predator and prey in space and time are important predation determining factors.

Growth in individual species is a complex process affected by many factors. First, it is dependent on the efficiency of digestion (*ie* how much of the eaten material is converted to tissue) and, thus, dependent on the organism itself and on some physical factors, such as temperature. As a population consists of individuals with different age and size, the growth rate of an individual is different than the average growth rate of the biomass. Furthermore, it depends also on the biomass of the growing organism present. Obviously the amounts of food available and taken are additional important growth determinants. Thus, growth is a complex dynamical process with many nonlinear terms.

Finally, we must consider the disposal (utilization) of growth and come by necessity to an 'inward' decreasing circulation system, in which the original unit amount of basic organic production changes from one organism to another, and decreases in quantity, but utilizes also other organic matter present in the system during the course of conversion. (See schematic *Fig 76* – 'the production spiral'.) The utilization encompasses also predation (including cannibalism) and includes other mortalities such as from diseases and old age. However, the organic matter does not end with the death of an organism – the carcasses are utilized to a large extent by many other biomasses, specially demersal fish, benthos, and ultimately bacteria. We can thus conclude that there is no definable surplus production in the marine ecosystem, but rather the system is a self perpetuating mechanism that is essentially in balance. Furthermore, there are no clearly defined food chains and trophic levels – these have been gross oversimplifications in the past.

Considering the above, we must also conclude that no single or simple theory, and no simple primitive equation formulations are possible to describe the marine ecosystem. This system must be presented and simulated with a set of equations of considerable size which reproduce individual processes and distributions. However, three important processes dominate quantitatively and control the abundance and distribution of individual components – growth, predation, and migration. Each of these processes is in turn controlled by a relatively complex set of conditions.

The most important direct food source (and food buffer) for marine fish is the zooplankton (which includes euphausiids and epibenthic crustaceans about which our present quantitative knowledge is deficient). The quantitative knowledge on zooplankton biomass, and especially about its production, is quite deficient in many areas and seasons. The reported average values by various authors can vary more than one order of magnitude in some regions. Furthermore, the pathways of plankton production through the food chain are extremely variable in space and time and quantitatively almost unknown. One of the most important factors in respect to the zooplankton availability to fish is its patchiness, about which our knowledge is, at best, qualitative. Thus, it is

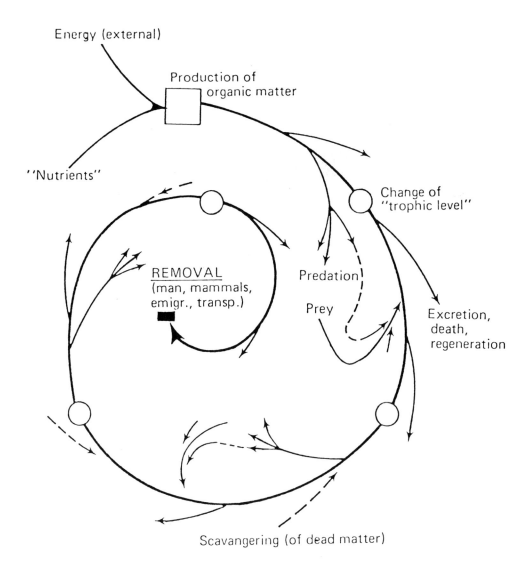

Energy (external)

Production of
organic matter

"Nutrients"

Change of
"trophic level"

REMOVAL
(man, mammals,
emigr., transp.)

Predation

Prey

Excretion,
death,
regeneration

Scavangering (of dead matter)

Fig 76   The biomass spiral.

virtually impossible to start any ecosystem model computations with basic organic and zooplankton production values. However, the zooplankton standing stock is simulated in the ecosystem computations, based on best available empirical data, where it buffers the limit of total finfish population to some extent.

The quantitative data on benthos are still more deficient than the data on plankton. Very little is known about the annual production of different components of benthos in many areas. Considering the very slow progress made in benthos and plankton research in the past fifty years, it appears unlikely that we could start a quantitative ecosystem model from primary production, zooplankton and benthos as 'producers'. It seems that all the quantitative computations which we can make at present in respect to benthos are of order of magnitude only.

For the purpose of the fish ecosystem simulation where benthos is the second 'production buffer', it is expedient to treat benthos in three ecological groups – predatory benthos, infauna, and epifauna.

Fish and other mobile marine organisms seem to be in constant search for food; by contrast most of the benthos and plankton have limited or no mobility at all. There seems to be a migration by fish into high food density areas and when the food is grazed down the grazers move into other areas, leaving former ('grazed down') areas for recovery. Thus, the zooplankton and benthos control the productivity in a large space and time scale and should be used as such in the ecosystem models.

11.2 Ecosystem simulation and the simulation of environmental effects

Summary
Numerical simulation of fish ecosystem is used for (*a*) evaluation of the resources by computations of the equilibrium biomasses, (*b*) simulation of the state of the ecosystem and natural fluctuations in it, and (*c*) simulation of the response of the ecosystem to fishery.

Three basic conditions of the ecosystem simulation are that it be based on empirical knowledge rather than theoretical conceptualization, that it includes all biota and biota-environment interactions, and that the simulation is conditionally unstable.

A biomass-based skeleton ecosystem model is described in this subchapter. The skeleton model allows the study and determination of some important environmental effects and permits also simulation of recruitment variations and migrations.

In this subchapter we present a brief outline of an ecosystem simulation approach which permits, to a limited extent, the study of the effects of environmental changes on the fish ecosystem. Details of this model are described by Laevastu and Larkins (1981). This ecosystem simulation is an 'upside down' model based on principles shown in *Fig 77*. In this simulation we are concerned with the dynamics of the biomasses of single species and/or groups of species. The factors and processes affecting these dynamics are schematically shown in *Fig 78*.

It might be useful to differentiate between various types of models and simulations. Conventionally, a model has been considered to be an abstraction and simplification of a given condition and/or process, whereas a simulation is a numerical reproduction of a system of conditions and processes, based on available empirical data and knowledge and may contain many tested models.

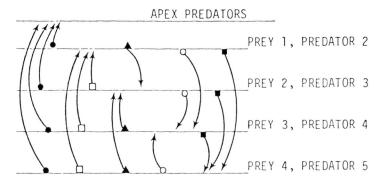

APEX PREDATORS

PREY 1, PREDATOR 2

PREY 2, PREDATOR 3

PREY 3, PREDATOR 4

PREY 4, PREDATOR 5

Principles: Determine who eats what and how much and then determine how much of the prey must be there to produce the eaten amounts.

Advantages: Minimum values of the production and standing stocks of all prey can be computed.

Amounts of noncommercial (and nonsampled) species can be estimated.

Changes in one prey biomass are related to changes in other prey biomasses.

Fig 77 Principles of trophodynamic computations, based on consumption.

The objectives of numerical ecosystem simulations can be grouped into two main categories:

(1) Investigative and digestive (analytical) objectives, including basic ecological research, that permit quantitative simulation of the state of the ecosystem, simulation of the effects of environmental changes and interspecies interactions in space and time, and the establishment of hypothesis and research priorities.

(2) General management guidance, the assessment of fisheries resources, and the effects of exploitation.

The following principles are normally followed in ecosystem simulation:

——The ecosystem simulation must include all of the essential biological and environmental interactive components of the system.

——Theoretical conceptualizations should be avoided, unless they have been tested with empirical data and proven to be valid.

——Biomass balance and trophodynamic computations should start with apex predators (including man); these can be treated as 'forcing functions' of the system.

——The system of equations should not be conditionally stable (except for unique solution in defined conditions).

164

A. Biomass abundance affecting processes

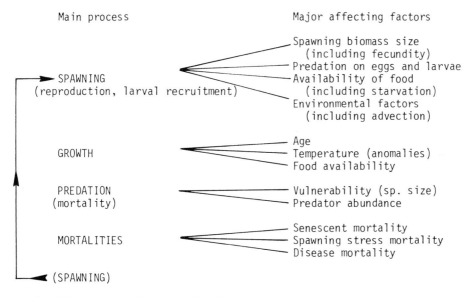

Main process   Major affecting factors

SPAWNING
(reproduction, larval recruitment)   Spawning biomass size
                                            (including fecundity)
                                          Predation on eggs and larvae
                                          Availability of food
                                            (including starvation)
                                          Environmental factors
                                            (including advection)

GROWTH   Age
                                          Temperature (anomalies)
                                          Food availability

PREDATION   Vulnerability (sp. size)
(mortality)   Predator abundance

MORTALITIES   Senescent mortality
                                          Spawning stress mortality
                                          Disease mortality

(SPAWNING)

B. Biomass distribution affecting processes

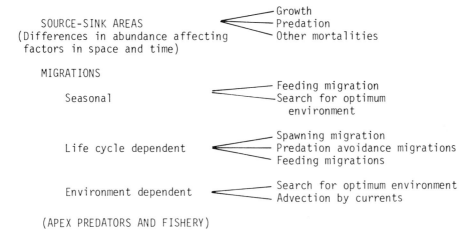

SOURCE-SINK AREAS   Growth
(Differences in abundance affecting   Predation
 factors in space and time)   Other mortalities

MIGRATIONS

    Seasonal   Feeding migration
                                          Search for optimum
                                            environment

    Life cycle dependent   Spawning migration
                                          Predation avoidance migrations
                                          Feeding migrations

    Environment dependent   Search for optimum environment
                                          Advection by currents

(APEX PREDATORS AND FISHERY)

Fig 78   Major dynamic processes in the marine ecosystem.

The skeleton model presented here is biomass-based (in contrast to conventional number-based models). The equations presented here can be applied to any fish species. To simplify the presentation, formulations and treatment of plankton and mammals (apex predators) are excluded. The biomass and trophodynamic equations can also be applied, with some modifications, to a single cohort of any species. The numerical behaviour of the individual formulas is well known and not described here.

165

The environmental effects can be studied effectively with an ecosystem simulation model which has three-dimensional space and a time dimension. Thus we use a computational grid (see example *Fig 79*) where computations are carried out at each grid point with the prevailing condition there. The third space dimension is depth, *eg* a surface layer grid and a bottom grid are used. Migrations (and advection) are simulated from grid point to grid point.

The biomass growth and mortality are computed in discrete time steps. The biomass growth rate is computed from empirical data of annual growth rates and distribution of biomass with age. The latter is computed with an auxiliary model (Granfeldt, 1979).

The biomass (B) of a cohort, species, or group of species (i) at the end of a given time step (t) (monthly time step is normally used) is computed with a well-known formula (16), using biomass from previous time step (t-1) and growth rate (coefficient) (g) minus total mortality rate (Z) for this time step.

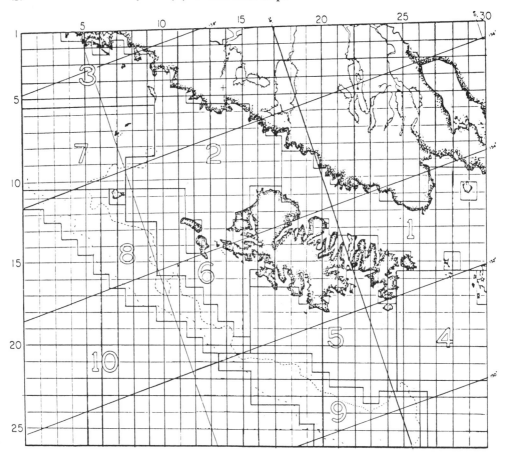

Fig 79   Computational grid with subregions for Kodiak area.

166

$$B_{i,t} = B_{i,(t-1)} e^{g_{i(t)} - Z_{i(t)}} \qquad (16)$$

The yield (Y) is computed with a prescribed fishing mortality coefficient ϕ_i. It should be noted that all the instantaneous coefficients (growth, mortality, fishery) are different than the corresponding conventional coefficients for number-based models which use annual time step. Thus all these coefficients have to be computed on biomass base and for the time step used in the model.

$$Y_{i,t} = B_{i,t} - B_{i,t} e^{-\phi_{i(t)}} \qquad (17)$$

The growth coefficient is computed in each time step, accounting for the effects of starvation in previous time step:

$$g_{i(t)} = g_i^o [(R_{i,t-1} - S_{i,t-1})/R_{i,t-1}] \qquad (18)$$

If there was no starvation in previous time step ($S_{i,t-1} = 0$), the rate of growth ($g_{i(t)}$) will take the prescribed value g_i^o, but if the species was not able to get all the food required for maximum growth rate ($R_{i(t)}$), the prescribed growth rate will be reduced by the ratio of the amount of food which the species was not able to get during the previous time step ($R_{i,t-1} - S_{i,t-1}$) over the total amount of food required by the biomass to grow under unlimited conditions ($R_{i,t-1}$). Both values are available from previous time step and the possible error caused by this necessary backstepping choice is again minimized by the use of short time step in the computations.

The effect of environmental temperature on growth is computed with the following formula:

$$g_{i(t)} = g_{i(t)} e^{\left(\frac{1}{T_0} - \frac{1}{T}\right)} \qquad (19)$$

where T_0 is acclimatization temperature for given species, and T is actual (monthly mean) temperature (either of surface layer or bottom, depending whether the species is pelagic or demersal).

The mortality rate ($Z_{i(t)}$) is the addition of all negative rates of changes representing thus the total mortality rate:

$$Z_{i(t)} = \phi_{i(t)} + \mu_i + \beta_{i,t-1} \qquad (20)$$

All rates of change are presented as instantaneous coefficients and are therefore additive. Fishing mortality ($\phi_{i(t)}$) and natural mortality from old age and diseases, including also spawning stress mortality (μ_i) are prescribed, but the predation mortality coefficient ($\beta_{i,t-1}$) is computed trophodynamically in previous time step from the ratio of consumption of the species over its biomass [$\beta_{i,t-1} = \ln\{1 - (C_{i,t-1}/B_{i,t-1})\}$].

167

The amount of food eaten by a species ($R_{i(t)}$) with unlimited food availability is:

$$R_{i(t)} = B_{i,t} \, r_i \, \tau \tag{21}$$

where r_i is the prescribed daily ration (in fraction of body weight daily) and τ is the length of time step in days. As temperature affects growth and metabolism, the daily ration must be made a function of temperature in the same manner as growth:

$$r_{i(t)} = r_i \, e^{\left(\frac{1}{T_0} - \frac{1}{T}\right)} \tag{22}$$

If the food supply of all food items for a given species was unlimited, we could compute the consumption of each food item [*eg* the consumption of species j by species i ($C_{j,i}$)] from the food requirement (R_i) and the fraction of species j (prey) in the food of species i (predator) ($\pi_{i,j}$):

$$C_{j,i} = R_{i,t} \, \pi_{i,j} \tag{23}$$

In this case the total consumption of species i would be:

$$C_i = \sum_j C_{i,j} \tag{24}$$

and the starvation would be 0. However, some food might be in limited supply and only part of the biomass of a prey is usually accessible as suitable food (*ie* size dependent feeding). The vulnerability of one species (prey) to another species (predator) is prescribed by average composition of the food of predator. Therefore, the fraction of each species which is allowed to be consumed in each time step is prescribed in the model (p_j), considering mainly the size composition of the biomasses of individual species. Furthermore, substitution of low-availability food items with high-availability items must be used. However, conditions can arise where full substitution is unrealistic and partial starvation will occur. There are various ways of computing the actual consumption with above described limitations.

The recruitment is usually depicted in number-based models as a discontinuity relating it to discrete spawning period. In our biomass-based model we have treated it as a continuous process. This treatment is acceptable if we think in terms of size groups rather than age groups and consider variations in growth of individuals belonging otherwise into the same age group, and assume a longer spawning period.

If the effect of environment on the recruitment of any species is quasi-quantitatively known, it should be included in the simulation model with proper empirical formula.

If the biomasses of all species in the ecosystem do not change over a year (*ie* previous January biomass is the same as actual January biomass), then we can say that the biomasses are in equilibrium. This implies that the growth of the biomass equals its removal by mortalities (specially by predation). If we want to achieve this equilibrium, we can change either growth rate, mortality rate, or biomass level itself. The growth rate is determined by empirical data and the other factors, such as temperature, are assumed in equilibrium case to be the same from one year to another (although seasonal changes can occur). Fishing and other mortality rates are also assumed to remain the same from one year to another. The predation mortality (consumption together with other mortalities which remain unchanged) must then balance the growth rate. This balancing can be achieved if the biomass levels of the predators are adjusted so that the biomasses remain constant from one January to another January. This adjustment can be done by finding a unique solution to the biomass equations of all species (or groups of species) in the ecosystem. This unique solution exists when one of the biomasses and con-sumption by it is predetermined (assumed to be known and fixed). In this case an iterative solution can be applied to adjust the biomasses of other species once after each year's computation:

$$B_{i,t_{12},0} = B_{i,t_{12},a} + \frac{(B_{i,b} - B_{i,a})}{k} \tag{25}$$

where $B_{i,t_{12},0}$ is the new (adjusted) biomass for December, $B_{i,t_{12},a}$ is the previous December biomass, $B_{i,b}$ is the biomass of previous January (computed as next step from $B_{i,t_{12},a}$), $B_{i,a}$ is the computed biomass in January one year later and k is an iteration constant (3·5 to 10, depending on the state of convergence). Forty years or more of simulation is needed before the solution converges to a unique (equilibrium) solution.

The model requires as input a number of species specific constants. Besides these, the biomass of at least one species must be prescribed as known (*ie* not altered in iterative adjustment). The biomasses of other species must be initially prescribed as the best first guesses. The first guess values of the consumption (C) can be computed by assuming C_i to be 8% of B_i per month.

In order to determine the carrying capacities of given ocean regions with the model and to obtain realistic equilibrium biomasses, the model must include all species. Computer capacity as well as basic information available does not usually allow the specification of all species separately, but many species must be grouped into ecological groups, whereby the composition of food and feeding habits are the main criteria for grouping.

Seasonal and life-cycle migration speeds may be prescribed as U and V components.

Migration computation is carried out in two steps. First, the linear gradients of biomass in the 'upcurrent' (upmigration) (UT and VT) are determined:

U positive:
$$UT_{(n,m)} = (B_{n,m} - B_{n,m-1})/l \tag{26}$$

U negative:
$$UT_{(n,m)} = (B_{n,m} - B_{n,m+1})/l \tag{27}$$

The computation of VT is analogous to UT computations above. In above formula l is grid length and n and m are the grid coordinates.

In the second step, the gradient is advected to the grid point under consideration.

$$B_{(t,n,m)} = B_{(t-1,n,m)} - (t_d \mid U_{(t,n,m)} \mid UT_{(n,m)}) - \qquad (28)$$
$$(t_d \mid V_{(t,n,m)} \mid VT_{(n,m)}) -$$

After each time step a smoothing (diffusion) operation is performed, which also simulates the random migration of fish:

$$B_{(n,m)} = \gamma B_{(n,m)} + \beta (B_{n-1,m} + B_{(n+1,m)} + B_{n,m+1} + B_{n,m-1}) \qquad (29)$$

where γ and β are smoothing coefficients ($\gamma = 0.8$ to 0.96 and $\beta = (1-\gamma)/4$).

The migrations due to unfavourable environment and/or search for food is done by checking surrounding grid points for various prescribed unfavourable–favourable criteria (and/or presence of optimum conditions) and corresponding to the success of this search a portion of the biomass is moved toward optimum conditions:

$$B_{i(n,m)} = B_{i(n,m)} - k_o B_{i(n,m)} \qquad (30)$$

$$B_{i(n \pm 1, \, m \pm 1)} = B_{i(n \pm 1, \, m \pm 1)} + k_s B_{i(n,m)} \qquad (31)$$

where: k_o is the fraction of biomass removed from the grid point and k_s is the fraction of this biomass added to a given neighbouring point. The size of these fractions depends on the nature of the 'forced migration' and on the number of favourable and unfavourable grid points surrounding the grid point under consideration.

Detailed description of the above simulation model is given by Laevastu and Larkins (1981), where results of the applications and the validation of the results are also described.

11.3 Modern fisheries management on the basis of ecosystem approach
Summary
The total biomass of fish in a given region changes little over time. However, the stocks of individual species can vary considerably, some species decreasing, others increasing in abundance. Any given combination of fishing results in different quantitative composition of ecosystem, whereby not only target species are affected by fishery, but also non-target species are influenced via interspecies interactions (mainly predation).

Thus, the ecosystem simulation estimates the response (behaviour) and state of the ecosystem resulting from different combinations of fishing (species and harvests) and environmental influences. The management regime has then to determine which of the resulting states of the ecosystem is acceptable from socio-economic and conservation points of view.

Some of the general principles of fisheries management are: (*a*) maintenance of sufficient reproductive potential to account for possible natural fluctuations, (*b*) maintenance of particular commercial species at economically harvestable levels, and (c) minimization of adverse effects on other potentially harvestable resources or otherwise desirable elements of the marine ecosystem.

One of the main objectives of the ecosystem simulation (modelling) is to compute quantitatively the response of the ecosystem to exploitation and to compute how much of any given species can be taken from the ecosystem of a given region without causing changes of undesirable nature and extent. The decision of what constitutes 'undesirable changes' in the ecosystem must be based on social, economic, and ecological criteria, some of which have little to do with traditional science. However, science must show the nature and extent of the changes and make suggestions as to which changes are acceptable from the ecosystem point of view.

The effects of fishing on the ecosystem can only be computed in a full (complete) ecosystem simulation model where all species and all fisheries are included. In such a complete system, removal of a given quantity of fish will cause changes in the biomass of the target species as well as in other species. The criteria for acceptable change in the ecosystem are usually established on regional bases with special consideration of the system's capacity to reproduce harvestable species biomasses. Such criteria will normally vary in time with changing economic, social, and ecological conditions.

Ecosystem simulations require as input fishing intensity coefficients for commercially exploited species. These coefficients are either:

(1) obtained from catch statistics,
(2) estimated on the bases of proposed regulations, or,
(3) estimated as possible acceptable catch using some other indices (see Laevastu and Larkins, 1981).

The first guess estimate of catch (fishing intensity) will merely serve as first guess input of fishing intensity (and/or yield) into the ecosystem in order to compute the response of the ecosystem to those yields and to provide background for management decisions. The first guess estimate will reduce the trial-and-error inputs for testing of the effects of fisheries with these models. The complex and complete simulation models also allow us to evaluate the changes in the stocks due to environmental change as well as those fluctuations in abundance caused by fishery.

The total finfish biomass in a given region fluctuates but little in the course of time (*Fig 80*), but individual members (species) can fluctuate considerably, one decreasing, another increasing. These individual species fluctuations are not always caused by fishing, but may be caused by other factors such as environmental anomalies. Thus, the determination of the causes, magnitudes, and periods of these 'natural fluctuations' is one of the important tasks in modern fisheries science.

The total finfish biomass (the carrying capacity of finfish) is determined by the

171

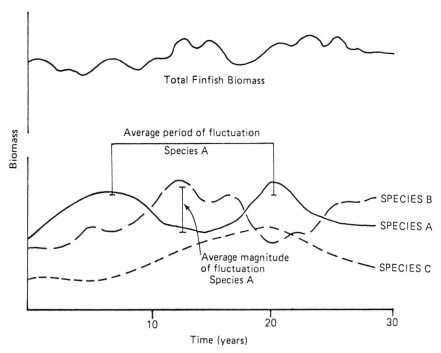

Fig 80 Schematic presentation of fluctuations of total finfish biomass and biomasses of individual species with time.

production of organic matters and its availability, especially in the form of food (benthos and zooplankton). The true carrying capacity is always smaller than the theoretical carrying capacity computed with the assumption of full utilization of organic production.

Figure 81 illustrates schematically what happens to biomass distribution with age when a virgin stock comes under the fishery. Within about five years the older part of the biomass (exploitable biomass) which was originally in balance with natural mortality (*ie* its distribution with age was determined by spawning stress mortality) will decrease until it is again in balance with the sum of fishing mortality and natural mortality. This reduction may result in faster growth of the remaining biomass and rejuvenation of the population, which has been observed in many fish stocks.

Three basic criteria are useful in establishing acceptable catch for any species:

(*1*) Maintenance of a reasonably high potential to reproduce in a commercially desirable species, *ie* to keep the biomass at a level where occasional 'recruitment failures' will not appreciably affect future recruitment. Thus, we must know the state of the resource (*ie* the level of the biomass) and consider in addition: age of maturity in relation to fishery (*ie* fully exploited year-class), spawning stress mortality, fecundity, time of fishery in relation to spawning period (assuming fishery can be regulated in space and time), the life span of the species, and a good idea of the magnitudes and periods of recruitment variations.

172

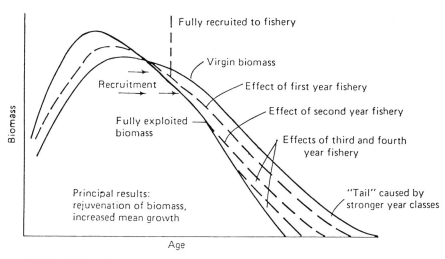

Fig 81 Schematic presentation of the effect of fishery on age composition of biomass, assuming recruitment and fishing intensity is constant.

(2) Minimization of any adverse effect on the other resources and ecosystem at large. This requires the knowledge of the ecosystem response to fishery which can be evaluated in large simulation models. One of the less considered aspects is the shrinking of the resource distribution (and spawning areas) with the decrease of biomass. Of concern also are the indirect effects of the fishery on other economically important species (other than 'target species'). Here the economic aspects enter into consideration, and the task of the biologist is to advise as to what might happen or what is expected to happen as a consequence of specific alteration decisions.

(3) The harvesting and economic aspects – *ie* maintaining the resource at a level where harvesting is profitable. Although some 'maximization' or 'optimization' of the production would be possible, it is seldom economically and politically possible – *ie* to achieve Maximum Sustainable Yield in an economic sense.

Additional consideration must be given to the following factors and processes which are expected to affect acceptable catch.

(1) Can the 'natural fluctuations' which occur in the ecosystem without the influence of fishing be separated from the effects of fishing ? What are the interactions between fishing and 'natural fluctuations' ? The magnitudes and periods of these 'natural fluctuations' caused by a variety of factors, such as temperature anomalies, may be simulated with ecosystem models before introducing changes in fishing.

(2) What is the process of 'recovery' of a stock, what are the factors determining it, and what are the 'recovery' speeds (relatively sudden or slow and gradual) ? Again, the recovery process may be investigated with a full ecosystem model.

173

(3) What is the state of a given stock in relation to equilibrium biomass ? At what level may it be expected to reach equilibrium ? How does this level compare to other possible equilibrium conditions ? Again, these questions may be investigated with a full scale ecosystem model.

During the iterative procedure in determining the changes in the ecosystem caused by different estimated catches, additional information will be produced such as:

(1) Quantitative changes taking place in principal predator-prey relations in the ecosystem.

(2) Changes in age distribution and recruitment to exploitable stock.

(3) Changes of growth rate and age of maturity.

(4) Changes in by-catch composition in mixed fishery.

The recommended acceptable catches or optimum yields can best be determined by management bodies after the plausible changes which will take place in the ecosystem are simulated in a complete ecosystem model and presented to the management body for evaluation and consideration. The management body, consisting of a variety of interests, will have then to decide which changes are acceptable in relation to different catches and exploitation strategy.

12

References

AHLSTRÖM, E H. Synopsis on the biology of the
1959   Pacific sardine (*Sardinops caerulea*).
        Scient. Meet. Biol. Sardines, Rome,
        1959.

ALEEV, IU G. Prisposoblenie k dvizheniiu i
1958   povorotlivost ryb (Adaptation to
        movement and manoeuvrability of fish).
        C. R. Acad. Sci. USSR, 120, 510–513.

ALLEN, E J. Reports on the present state of
1897   knowledge with regard to the habits and
        migrations of the mackerel (*Scomber
        scombrus*). *J. Mar. Biol. Ass. UK*,
        5,1–40.

ALLEN, E J. Mackerel and sunshine. *J. Mar.*
1909   *Biol. Ass.* UK, 8, 394–406.

ALVERSON, D L. A study of annual and
1960   seasonal bathymetric catch patterns for
        commercially important groundfishes of
        the Pacific North-west coast of North
        America. *Bull. Pac. Mar. Fish. Comm.*
        Portland, 4, 66 pp.

ALVERSON, D L, PRUTER, A T and RONHOLT,
1964   L L. A study of demersal fishes and
        fisheries of the Northeastern Pacific
        Ocean. *Inst. of Fisheries*, Univ. of British
        Columbia, Vancouver, 190 pp.

ANDERSEN, K P and URSIN, E. A multispecies
1977   extension to the Beverton and Holt
        theory of fishing, with accounts of
        phosphorus circulation and primary
        production. *Medd. Danm, Fish.
        Havunder.* (Ny Ser) 7:319–435.

ANTONY RAJA, B T. Possible explanation for
1974   the fluctuation in abundance of the
        Indian oil sardine, *Sardinella longiceps*,
        Valenciennes. *Proc. IPFC*, 15(3):241–
        52.

BALCHEN, J G. Modeling, predation and
1979   control of fish behavior. *In* Control and
        Dynamic Systems. Academic Press,
        pp 100–146.

BALLS, R. Environmental changes in herring
1951   behaviour: A theory of light avoidance
        as suggested by echo-sounding
        observations in the North Sea. *J. Cons.
        ICES*, 17(3), 274–298.

BARKLEY, R A. Salinity maxima and the
1969   skipjack tuna *Katsuwonus pelamis*. *Bull.
        Jap. Soc. Fish. Oceanogr.*, Spec. Publ.
        No:243–8.

BAXTER, J L. Summary of biological infor-
1967   mation on the northern anchovy
        (*Engraulis mordax*, Girard). *Cal. Coop.
        Oc. Fish. Invest. Rpt.* 11, 110–116.

BELL, F H and PRUTER, A T. Climatic
1958   temperature changes and commercial
        yields of some marine fisheries. *J. Fish.
        Res. Bd. Canada*, 15(4):625–683.

BERZINS, B. Ueber temperaturbedingte
1949   Tierwanderungen in der Ostsee.
        Oikos, 1, 29–33.

BEVERTON, R J H and LEE, A J. The influence
1965   of hydrographic and other factors on
        the distribution of cod on the Spitz-
        bergen shelf. *ICNAF Spec. Publ.*, 6,
        225–245.

BISHAI, H M. The effect of water currents on
1960   the survival and distribution of fish
        larvae. *J. Cons. ICES.*, 25,(2), 134–146.

BITIUKOV, E G. Kvoprosu o sutochnykh
1959   vertikal'nykh migratsiakh salaki. *Dokl.
        Aka. Nauk SSSR*, 128(1), 179–182.

BLAXTER, J H S. The effects of extremes of
1960   temperature on herring larvae. *J. Mar.
        Biol. Ass. UK*, 39(3), 605–608.

BLAXTER, J H S. Effect of change of light
1965   intensity on fish. *ICNAF Spec. Publ.*,
        6, 647–661.

175

BLAXTER, J H S and DICKSON, W. Observations
1959   on the swimming speeds of fish. *J.
       Cons. ICES*, 24(3), 472–479.
BLAXTER, J H S, HOLLIDAY, F G T and
1958   PARRISH, B B. Some preliminary
       observations on the avoidance of
       obstacles by herring (*Clupea harengus*
       L.). *Symp. Pap. IPFC*, 7, 5 pp.
BLAXTER, J H S and PARRISH, B B. The effect
1958   of artificial lights on fish and other
       marine organisms at sea. *J. Mar. Res.*,
       2, 24 pp.
BRATBERG, E and HYLEN, A. A study of the
1964   relationship between the water
       temperature and the concentration of
       cod in West Greenland waters. *Rpt.
       Norw. Fish. and Mar. Inves.*, 13(7),
       17–26.
BRAWN, V M. Seasonal and diurnal vertical
1960   distribution of herring (*Clupea harengus*
       L.) in Passamaquoddy Bay, N.B. *J. Fish.
       Res. Bd. Canada*, 17(5), 699–711.
BRETT, J R. Salmon research and hydroelectric
1957   power development. *Bull. Fish. Res.
       Bd. Canada*, 114, 1–26.
BRETT, J R, HOLLANDS, M and ALDERDICE, D
1958   F. The effect of temperature on the
       cruising speed of young sockeye and
       coho salmon. *J. Fish. Res. Bd. Canada*,
       15, 587–605.
BRIDGER, J P. On day and night variation in
1956   catches of fish larvae. *J. Cons. ICES*,
       22(1), 42–57.
BULL, H O. An evaluation of our knowledge of
1952   fish behaviour in relation to hydro-
       graphy. *Rapp. ICES*, 131, 8–23.
BUYS, M E L. The South African pilchard
1959   (*Sardinops ocellata*) and maasbanker
       (*Trachurus trachurus*) – Hydrographical
       environment and the commercial
       catches, 1950–57. *Invest. Rep. Div. Fish.
       S. Afr.*, 37, 559–573.
CARRUTHERS, J N. Fluctuations in the herrings
1938   of the East Anglian autumn fishery, the
       yield of the Ostend spent herring fishery
       and the haddock of the North Sea in
       the light of the relevant wind con-
       ditions. *Rapp. P.-v Réun. ICES*,
       107:1–15.
CARRUTHERS, J N. Fish, fisheries and environ-
1956   mental factors. *Oceanus*, 4(2), 14–20.
CARRUTHERS, J N, LAWFORD, A D and VELEY,
1951   V F C. Fishery hydrography: brood-
       strength fluctuations in various North

Sea fish, with suggested methods of
prediction. *Kiel. Meeresforsch.*,8(1),5–15.
CHAPMAN, W M. The application of oceano-
1972   graphy to the development and manage-
       ment of ocean fisheries. *Publ. Fish.
       Univ. Wash.*, 5:13–21.
CHASE, J. Winds and temperatures in relation
1955   to the breed-strength of Georges Bank
       haddock. *J. Cons. ICES*, 21, 17–24.
CHASE, J. Wind-induced changes in the water
1959   column along the East Coast of the
       United States. *J. Geophys. Res.*, 64(8),
       1013–1022.
CHERNYAVSKIY, V I. Relation between catch of
1970   fattening Okhotsk herring and water
       temperature. *Izv. Tikhookean.
       Nauchno-Issled. Inst. Rybn. Khoz.
       Okeanogr.*, 71:51–8 (in Russian).
CLARKE, G L. On the depth at which fish can
1939   see. *Ecology*, 17, 452–456.
CLAY, C S and MEDWIN, H. Acoustical
1977   oceanography. John Wiley and Sons,
       New York. 544 pp.
COLTON, J B, Jr. Temperature trends and the
1972   distribution of groundfish in continental
       shelf waters, Nova Scotia to Long
       Island. *Fish. Bull. NMFS/NOAA*,
       70(3):637–57.
CORLETT, J. Distribution of larval cod in the
1958   Western Barents Sea. *ICNAF Spec.
       Meet. Biarritz* (Mimeo).
CORLETT, J. Winds, currents, plankton and the
1965   year-class strength of cod in the
       Western Barents Sea. *ICNAF Spec.
       Publ.*, 6, 373–378.
CRAIG, R E. A relation between herring drifter
1957   catches and hydrographic conditions.
       ICES, C. M. 1957. Herring Committee,
       No. 58, (Mimeo).
CRAIG, R E. Dependence of catches on
1958   temperature and wind in the Buchanan
       spawning fishery. *ICES, C. M. 1958,
       Herring Committee*, No. 43 (Mimeo).
CUSHING, D H. The detection of fish.
1973   Pergamon Press, Oxford. 200 pp.
CUSHING, D H. Recruitment and parent stock
1976   in fisheries. *Univ. Wash., Div. Mar.
       Resour., Wash. Sea Grant Program*,
       WSG 73–1, 197 pp.
CUSHING, D H and DICKSON, R R. The
1976   biological response in the sea to
       climatic changes. *Adv. Mar. Biol.*,
       London, Academic Press, 14:1–122.
DAAN, N. A quantitative analysis of the food

1973   intake of North Sea cod, *Gadus morhua.*
       Neth. J. Sea Res. 6(4):479–517.

DANNEVIG, A. Mackerel and sea temperature,
1955   measurements – 21 April to 15 May,
       1952. Praktiske Fiskeforsok, 1952.
       Arsber. Norges Fisk., 5, 64–67.

DANNEVIG, H. The influence of temperature on
1895   the development of the eggs of fishes.
       Ann. Rep. Fish. Bd., Scotland 1894,
       13(3) (5), 147–152.

DAVIDSON, V M. Salmon and eel movement in
1949   constant circular current. *J. Fish. Res.
       Bd., Canada,* 7, 432–448.

DAVIES, D H. The South African pilchard
1956   (*Sardinops ocellata*). Migration, 1950–55.
       *Invest. Rep. Fish. Mar. biol. Surv. S.
       Afr.,* 24, 1–52.

DE CIECHOMSKI, J DZ. Influence of some
1967   environmental factors upon the
       embryonic development of the
       Argentine anchovy, *Engraulis anchoita*
       (Hubbs, Marini). *Cal. Coop. Oc. Fish.
       Invest. Rpt.,* 11, 67–71.

DEVOLD, F. Herring trip with *G O Sars* in
1951   Norwegian Sea 5/7–24/8, 1950.
       Praktiske Fiskeforsok, 1950. *Arsberet.
       Norg. Fisk.,* 5, 110–119.

DEVOLD, F. Otto Petterssons teori om de
1959   skandinaviske sildeperioder sett i
       sammenheng med de senere aar
       undersokelser over den atlanto-
       skandinaviske sild. *Naturen, Bergens
       Mus.,* 2, 83–92.

DEVOLD, F. Report of the joint Icelandic-
1969   Norwegian-Soviet investigations on
       adult Atlanto-Scandian herring in
       summer 1969. *Fiskets Gang.,* 36:602–4.

DICKSON, R R, POPE, J G and HOLDEN, M J.
1974   Environmental influences on the
       survival of North Sea cod. *In* The early
       life history of fish. Blaxter J H S, Ed.
       Proceedings of an International
       Symposium held at the Dunstaffnage
       Marine Research Laboratory of the
       Scottish Marine Biological Association
       at Oban, Scotland, May 17–23, 1973.
       New York, Springer-Verlag, pp. 69–80.

DIETRICH, G, SAHRHAGE, D and SCHUBERT, K.
1959   The localization of fish concentrations
       by thermometric methods. *In* Modern
       fishing gear of the world. London.
       Fishing News (Books) Ltd., pp. 453–461.

DODIMEAD, A J and PICKARD, G L. Annual
1967   changes in the oceanic-coastal waters of

the Eastern Sub-arctic Pacific. *J. Fish.
       Res. Bd. Canada,* 24(11), 2207–2227.

DOUDOROFF, H P. The resistance and
1942   acclimatization of marine fishes to
       temperature changes, I. Experiments
       with *Girella nigricans* (Ayres). *Biol.
       Bull., Woods Hole,* 2, 219–244.

DOW, R L. A comparison among selected
1964   marine species of an association between
       sea water temperature and relative
       abundance. *J. Cons. ICES,* 28(3):
       425–31.

DOW, R L. Cyclic and geographic trends in sea
1969   water temperature and abundance in
       American lobster catches. *Science,
       Wash.,* 164:1060–2.

DRAGESUND, O. Reactions of fish to artificial
1958   light, with special reference to large
       herring and spring herring in Norway.
       J. Cons. ICES, 23, 213–227.

DRAGESUND, O and NAKKEN, O. Relationship of
1973   parent stock size and year-class strength
       in Norwegian spring spawning herring.
       Rapp. P.-v. Réun. ICES., 164:2–29.

DRAGESUND, O and OLSEN, S. On the possibility
1965   of estimating year-class strength by
       measuring echo-abundance of O-group
       fish. *Rpt. Norw. Fish. and Mar. Invest.,*
       13(8), 48–62.

EARLL, R E. A report on the history and present
1880   conditions of the shore cod fisheries of
       Cape Ann, Massachusetts, together
       with notes on the natural and artificial
       propagation of the species. *Rep. U.S.
       Fish. Comm., 1878,* 6, 685–740.

EGGVIN, J. Tilstanden i havet under den
1963   unormale vinter 1963. *Fisken og Havet,*
       1 (1963), 9–16.

EGGVIN, J. Pilot project on rapid utilization of
1966   synoptic oceanographic observations.
       ICES, CM. 1966, Hydrogr. Comm.
       No. 17 (Mimeo).

ELKIN, E YA. The problem of forecasting the
1973   time of formation of large schools of
       foraging Okhotsk herring. *Izv.
       Tikhookean. Nauchno-Issled. Inst.
       Rybn. Khoz. Okeanogr.,* 86:22–5 (in
       Russian). Issued also as Transl. *Ser.
       Fish. Mar. Serv. Can.,* (3243):7 pp.

ELLIS, G H. Observations on the shoaling
1956   behaviour of cod (*Gadus callarias*) in
       deep water relative to daylight. *J. Mar.
       Biol. Ass. U.K.* 35(2), 415–418.

ELSON, P H. Effects of currents on the move-

1939 ment of speckled trout. *J. Fish. Res. Bd. Canada*, 4(5):491–499.

FAO, FISHERIES DIVISION. Report of ACMRR
1967 working party on direct and speedier estimation of fish abundance. *FAO Fish. Rpt. 41, Suppl. 1* (Mimeo).

FAVORITE, F and LAEVASTU, T. A study of the
1979 ocean migrations of sockeye salmon and estimation of the carrying-capacity of the North Pacific Ocean, using a dynamical numerical salmon ecosystem model (NOPASA). *NW and Alaska Fish. Cent., Seattle, Proc. Rep.* 47 pp.

FAVORITE, F, LAEVASTU, T and STRATY, R R.
1977 Oceanography of the northeastern Pacific Ocean and eastern Bering Sea, and relations to various living marine resources. *NW and Alaska Fish. Cent., Seattle, Proc. Rpt.*, 280 pp.

FLITTNER, G A. Review of the movement
1964 of albacore tuna off the Pacific coast in 1963. *US Fish and Wildlife Serv., Comm. Fish. Rev.*, 26(12), 13–19.

FLITTNER, G A. Forecasting availability of
1970 albacore tuna in the Eastern Pacific Ocean. *In* Fisheries Oceanography. London, Fishing News (Books) Ltd: 116–129.

FLOWERS, J M and SAILA, S G. An analysis of
1972 temperature effects on the inshore lobster fishery. *J. Fish. Res. Board Can.*, 29(8):1221–5.

FORBES, S T and NAKKEN O (Eds.) Manual of
1972 methods for fisheries resource survey and appraisal. Part 2. The use of acoustic instruments for fish detection and abundance estimation. *FAO Mar. Fish. Sci.* 5, 138 pp.

FRANCIS, R C. Fisheries science now and in
1980 the future, a personal view. *N. Z. J. Mar. Freshw. Res.*, 14(1):95–100.

FRASER, J H. The drift of the planktonic stages
1958 of fish in the Northeast Atlantic and its possible significance to the stocks of commercial fish. *ICNAF, Spec. Publ.*, 1, 289–310.

GALLOWAY, J C. Lethal effect of the cold
1941 winter of 1939-1940 on marine fishes at Key West, Florida. *Copeia* 2, 128–229.

GALTSOFF, P S. Seasonal migrations of mackerel
1924 in the Black Sea. *Ecology* 5, 1–5.

GARROD, D J and COLEBROOK, J M. Biological
1978 effects of variability in the North Atlantic Ocean. *Rapp. P.-v Réun.*

ICES, 173:128–144.

GLOVER, R S, ROBINSON, G A and COLEBROOK,
1974 J M. Marine biological surveillance. *Environment and Change*, 2:395–402.

GRAHAM, J J. Central North Pacific albacore
1957 surveys, May to November 1955. *US Fish Wildl. Serv. Spec. Scient. Rep.*, 212, 1–32.

GRAINGER, R J R. Herring abundance of the
1978 west of Ireland in relation to oceanographic variation. *J. Cons. ICES*, 38(2):180–188.

GRANFELDT, E. Numerical method for
1979 estimation of fish biomass parameters. *NW and Alaska Fish Cent., Seattle, Proc. Rep.* 19 pp.

GREER WALKER, M, HARDEN JONES, F R and
1978 ARNOLD, G P. The movement of plaice (*Pleuronectes platessa L.*) tracked in the open sea. *J. Cons. ICES*, 38(1):58–86.

GRIFFITHS, R C. A Study of ocean fronts off
1965 Cape San Lucas, Lower California. *US Fish. Wildl. Serv. Spec. Scient. Rpt. Fisheries*, 499, 54 pp.

GUIEYSSE, L and SABATHE, P. Submarine
1964 Acoustics. (*Trans*) FNWF Monterey, CA.

HACHEY, H B, HERMANN, F and BAILEY, W B.
1954 The waters of the ICNAF Convention area. *Annual Proc. ICNAF*, 4, 67–102.

HANSEN, W. The reproduction of the motion
1966 in the sea by means of hydrodynamical-numerical methods. *Inst. Meereskunde, Univ. Hamburg, Rpt.* 5, 57 pp and figures.

HANZAWA, M, OKABAVASHI, K, YOSHIDA, K and
1951 MARUMO, R. Report on sea and weather observations on the Antarctic whaling ground, 1950-51. *Oceanogr. Mag.*, 3, 1–7–137.

HARDEN JONES, F R. Movements of herring
1957 shoals in relation to the tidal current. *Journ. Cons. ICES*, 22(3):322–328.

HARDEN JONES, F R. Fish migration. London,
1968 Ed. Arnold Ltd., 325 pp.

HARDEN JONES, F R and SCHOLES, P. Wind and
1980 the catch of a Lowestoft trawler. *J. Cons. ICES*, 39(1):53–69.

HARDER, W and HEMPEL, G. Studien zur
1954 Tagesperiodik der Aktivität von Fischen. I. Versuche an Plattfischen. *Kurze Mitteil. Inst. Fischereibiol. Univ. Hamburg*, 5, 22–31.

HASLER, A D, MORRALL, R H, WISBY, W J and
1958 BRAEMER, W. Sun-orientation and

homing in fishes. *Limnol. Oceanogr.*, 3, 353–361.

HAYASAKA, I. On the fatal effect of cold
1934   weather upon certain fishes of the sea around the islands of Hoko (the Pescadores Islands). A palaeontological point of view. *Taikohu Imp. Univ. Mem.*, 13(2), Geology, No. 9, 5–12.

HELA, I. The surface current field in the
1954   western part of the north Atlantic. *Bull. Mar. Sci. Gulf Caribb.*, 3(4), pp. 241–272, 1954.

HELA, I. The influence of temperature on the
1960   behaviour of fish. *Arch. Soc. 'Vanamo'*, 15, 1–2.

HELA, I and LAEVASTU, T. Fisheries hydro-
1962   graphy. London, Fishing News (Books) Ltd., 137 pp.

HELLAND-HANSEN, B and NANSEN, F.
1920   Temperature variations in the North Atlantic Ocean and in the atmosphere. Introductory studies on the causes of climatological variations. *Smiths. Misc. Coll.*, 70(4), 1–408.

HEMPEL, G. Studien zur Tagesperiodik der
1956   Aktivität von Fischen. II. Die Nahrungsaufnahme der Scholle. *Kurze Mitteil. Inst. Fischereibiol. Univ. Hamburg*, 6, 22–37.

HEMPEL, G. Auswirkungen des Wetterges-
1960   chehens auf die Fischbestände im Meer. *Umschau*, 60 (15), 466–469.

HEMPEL, G. Synopsis of the symposium on
1978a North Sea fish stocks – recent changes and their causes. *Rapp. P.-v. Réun. ICES*, 172:445–449.

HEMPEL, G. North Sea fisheries and fish
1978b stocks – a review of recent changes. *Rapp. P.-v. Réun. ICES*, 173:145–167.

HERMANN, F. Hydrographic conditions off
1951   the West Coast of Greenland, 1950. (With remarks on the influence of temperature on cod year-classes). *Ann. biol. ICES*, 7, 21–24.

HERMANN, F and HANSEN, P M. Possible
1965   influence of water temperature on the growth of the West Greenland Cod. *ICNAF Spec. Publ.*, 6, 557–563.

HERMANN, F, HANSEN, P M and HORSTEAD,
1965   S A. The effect of temperature and currents on the distribution and survival of cod larvae at West Greenland. *Spec. Publ. ICNAF*. No. 6, 389–409.

HESTER, F J. Identification of biological sonar

HIDA, T S. Chaetognaths and pteropods as
1957   biological indicators in the North Pacific. *US Fish Wildl. Serv. Spec. Scient. Rep.* 215, 1–13.

HODDER, V M. The possible effects of tem-
1965   perature on the fecundity of Grand Bank haddock. *ICNAF Spec. Publ.*, 6, 515–522.

HOLDEN, M J. Long-term changes in landings
1978   of fish from the North Sea. *Rapp. P.-v. Réun. ICES*, 172:11–26.

HOLLIDAY, F T T and BLAXTER, J H S. The
1960   effects of salinity on the developing eggs and larvae of the herring. *J. Mar. Biol. Ass. UK*, 39(3), 591–603.

HORSTED, SV AA and SMIDT, E. Influence of
1965   cold water on fish and prawn in West Greenland. *ICNAF Spec. Publ.*, 6, 199–207.

HORWOOD, J W and CUSHING, D H. Spatial
1978   distributions and ecology of pelagic fish. *In:* Spatial pattern in plankton communities (J H Steele, Ed.). NY, Plenum Press, 355–383.

HSIAO, S C. Reaction of tuna to stimuli – 1951,
1952   Part III. Observations on the reactions of tuna to artificial light. *US Fish and Wildlife Serv. Spec. Sci. Rept. Fish.*, 91, 36–58.

HUBERT, W E. Operational forecasts of sea and
1964   swell. *Proc. 1st US Navy Symp. on Milit. Oceanogr.*, 113–124.

HUBERT, W E. Computer produced synoptic
1964   analyses of surface currents and their application for navigation. *J. Inst. Navig.*, 12(2), 101–107.

HUBERT, W E and LAEVASTU, T. Synoptic
1965   analyses of forecasting of surface currents. *Fleet Numerical Weather Facility Technical Note*, No. 9.

HUNTER, J R. Effects of light on schooling and
1968   feeding of jack mackerel (*Trachurus symmetricus*). *J. Fish. Res. Bd. Canada*, 25 (2), 393–407.

HUNTER, J R and MITCHELL, C T. Field
1968   experiments on the attraction of pelagic fish to floating objects. *J. Cons. ICES*, 31(3):427–434.

HUTCHINS, L W. The bases for temperature
1947   zonation in geographical distribution. *Ecol. Monogr.*, 17(3):325–335.

HYLEN, A and DRAGESUND, O. Recruitment of
1973   young Arcto-Norwegian cod and
        haddock in relation to parent stock size.
        Rapp. P.-v. Réun. ICES, 164:60–68.

IMAMURA, Y. Study on the disposition of fish
1958   towards the light. II. The strength of
        illumination preferred by fish. *J. Tokyo
        Univ. Fish.*, 44($\frac{1}{2}$), 75–89.

INOUE, M and OGURA, M. The swimming-
1958   water-depth for anchovy shoals in
        Tokyo Bay. *Bull. Jap. Soc. Sci.
        Fish.*, 24, 311–316.

INOUE, M, NISHIZAWA, S, TAMUKAI, K and
1958   KUDO, T. Studies on the visual range of
        net twines in water. On the relation
        between the visual range of net twines
        and the turbidity of surrounding water.
        Bull. Jap. Soc. Sci. Fish., 24, 501–506.

ITO, S. On the relation between water
1958   temperature and the incubation time of
        sardine (*Sardinops melanosticta*). *Ann.
        Rep. Jap. Sea Reg. Fish. Res. Lab.*, 4,
        25–31.

JACKMAN, L A J and STEVEN, G A. Tem-
1955   peratures and mackerel movements in
        the inshore waters of Torbay, Devon-
        shire. *J. Cons. ICES*, 21(1), 65–71.

JAKOBSSON, J. The Icelandic herring search and
1971   information services. *In* Modern fishing
        gear of the world, Kristjonsson, H J ed.,
        London, Fishing News (Books) Ltd.,
        Vol. 3:2–11.

JAKOBSSON, J. The north Icelandic herring
1978   fishery and environmental conditions
        1960-1968. Cons. Int. explor. Mer,
        Symp. on the Biol. Basis of Pel. Fish
        Stock Management. Paper 30.

JAMES, R W. Application of wave forecasts to
1957   marine navigation. *US Navy Hydro-
        graphic Office, Spec. Publ.*, 1, 78 pp.

JEAN, Y. A study of spring and fall spawning
1956   herring (*Clupea harengus* L.) at Grande-
        Rivière, Bay of Chaleur, Quebec.
        Contr. Dept. Fish, 49, 1–76.

JENSEN, A J C. Account of the present know-
1955   ledge of the reaction of the mackerel to
        environmental factors. UN Resource
        Conf., Rome.

JOHANSEN, A C and KROGH, A. The influence of
1914   temperature and certain other factors
        upon the rate of development of the
        eggs of fishes. *Publ. Circ. ICES*, 68,
        43 pp.

JONES, R. Density dependent mortality of the

1973   numbers of cod and haddock. *Rapp.
        P.-v. Réun. ICES*, 164: 156–73.

JONSSON, J. Temperature and growth of cod in
1965   Icelandic Waters. *ICNAF Spec. Publ.*,
        6, 537–539.

KAMIURA, F. Mixing status of several fish
1958   species as revealed by fish school
        research. *Rep. Nankia Reg. Fish.
        Res. Lab.*, 7, 30–36.

KANDA, K, KOIKE, A and OGURA, M. The study
1958   on the colour of fishing net. III. Effect
        of the depth of colour of a net on the
        behaviour of a fish school near the net.
        Bull. Jap. Soc. Sci. Fish, 23,
        621–624.

KÄNDLER, R. An account of the knowledge
1955   concerning the response of the plaice to
        environmental factors. UN Resource
        Conf., Rome.

KAWAMOTO, N Y. Experiments with the fish
1958   gathering lamp. *Proc. Indo-Pacif.
        Fish. Coun.*, 6, Sec., 2–3, 278–280.

KAWAMOTO, N Y. The significance of the
1959   quality of light for the attraction of fish.
        In: Modern fishing gear of the world.
        London, Fishing News (Books) Ltd.,
        553–555.

KAWANA, T. On the relations between the
1934   oceanographical conditions and tuna
        fishing. (In Japanese). *J. Imp. Fish.
        Inst.*, 31, 1–80.

KETCHEN, K S. Preliminary experiment to
1957   determine the working of trawling gear.
        *Prog. rep. Pac. Coast Stat. Fish. Res.
        Bd., Canada*, 88, 62–65.

KOJIMA, S. Reactions of fish to a shade of
1957   floating substances. *Bull. Jap. Soc. sci.
        Fish.*, 22(12):730–735.

KOMAROVA, T V. Feeding of the long rough
1939   dab in the Barents Sea in connection
        with food resources. *Trans. Inst. Mar.
        Fish. Oceanogr. USSR*, 4, 298–320.

KONDO, H, HIRANO, Y, NAKAYAMA, N and
1965   MIYAKE, M. Offshore distribution and
        migration of Pacific salmon (genus
        Oncorhynchus) based on tagging studies
        (1958-1961). *Int. N. Pac. Fish. Comm.
        Bull. 17.*, 213 pp.

KONSTANTINOV, K G. Forecasting for the
1967   distribution of fish concentrations in
        the Barents Sea by a temperature factor.
        *Tr. Polyarn. Nauchno-Issled. Procktn.
        Inst. Morsk. Rybn. Khoz. Okeanogr.*,
        20:167–78 (in Russian).

KONSTANTINOV, K G. Importance of oceano-
1973 graphic data for development of fishery
forecasts. *Tr. Polyarn. Nauchno-
Issled. Procktn. Inst. Morsk. Rybn.
Khoz. Okeanogr.*, 34:67–72 (in Russian).

KONSTANTINOV, K G and SVETLOV, I I. On
1974 temporal relationship between water
temperature and the distribution of
marine commercial fishes. *Oceanology*,
14(2):349–54.

KURITA, S. Causes of fluctuations in the
1959 sardine population off Japan. Scienti.
Meet. Sardines, Rome 1959.

LAEVASTU, T. Factors affecting the temperature
1960 of the surface layer of the sea. *Soc.
Sci. Fenn., Comm. Phys.-Math.*,
25(1), 136 pp.

LAEVASTU, T. Natural bases for the fishery of
1961 the Atlantic Ocean, its present
characteristics and future possibilities.
In: Atlantic ocean fisheries. London,
Fishing News (Books) Ltd.

LAEVASTU, T. The adequacy of plankton
1962 sampling. *Rapp. ICES*, 153, 66–73.

LAEVASTU, T. Numerical oceanographic fore-
1969 casting for fisheries. *Bull. Jap. Soc.
Fish. Oceanogr., Spec. No:* 139–47.

LAEVASTU, T, FAVORITE, F and LARKINS, H A.
1979 Resource assessment and evaluation of
the dynamics of the fisheries resources
in the NE Pacific with numerical
ecosystem models. *NW and Alaska
Fish. Cent., Seattle, Proc. Rep.* 35 pp.

LAEVASTU, T and FAVORITE, F. Holistic
1980 simulation models of shelf seas
ecosystems. *In:* Analyses of marine
ecosystems (Longhurst, A. Ed.).
London, Academic Press. 701–727.

LAEVASTU, T and HELA, I. Fisheries
1970 oceanography. London, Fishing News
(Books) Ltd., 238 pp.

LAEVASTU, T and HUBERT, W E. Analysis and
1965 prediction of the depth of the thermocline and near-surface thermal structure.
FNWC Techn. Note, No. 10.

LAEVASTU, T and JOHNSON, J. Application of
1971 oceanographic and meteorological
analyses/forecasts in fisheries. *In*
Modern fishing gear of the world.
Kristjonsson, H J. Ed. London, Fishing
News (Books) Ltd., vol. 3:28–40.

LAEVASTU, T and LARKINS, H A (in press).
1981 Marine fisheries ecosystem: its
quantitative evaluation and

management. Farnham, Fishing News
Books Ltd.

LAEVASTU, T and ROSA, H Jr. The distribution
1963 and relative abundance of tunas in
relation to their environment. *FAO
Fish. Rep.* (6) 3:1835–51.

LAFOND, E C. Oceanography and food.
1961 *Naval Res. Rev.*, Nov., 1961, 9–13.

LAFOND, E C. Detailed temperature structures
1963 of the sea off Baja California. *Limnol.
Oceanogr.*, 8(4), 417–25.

LARKIN, P A. An epitaph for the concept of
1977 maximum sustained yield. *Trans. Am.
Fish. Soc.* 106 (1):1–11.

LAURS, R M and LYNN, R J. Seasonal migration
1978 of North Pacific albacore, *Thunnus
alalunga*, into North American coastal
waters: distribution, relative abundance
and association with transition zone
waters. *Fish. Bull. NMFS/NOAA.*

LAUZIER, L. Recent water temperatures along
1952 the Canadian Atlantic coast. *Progr.
Rep. Atl. biol. Sta.*, 53, 5–7.

LAUZIER, L. Effect of storms on the water
1957 conditions in the Magdalen Shallows.
Bull. Fish. Res. Bd., Canada III, 185–
192.

LE DANOIS, E. History of the investigations
1934 made at Newfoundland. *Proc. North
Amer. Counc. Fish. Invest.*, 1:35–56.

LEE, A J. The influence of hydrography on the
1952 Bear Island cod fishery. *Rapp. ICES*,
131, 74–102.

LEE, A J. British fishery research in the
1956 Barents Sea. *Polar Res.*, 8(53), 109–117.

LEE, A J. The influence of environmental
1959 factors on the Arcto-Norwegian cod
stocks. *Prepr. Int. Oceanogr. Congr.,
New York*, 243–245.

LOESCH, H. Sporadic mass shoreward
1960 migrations of demersal fish and
crustaceans in Mobile Bay, Alabama.
Ecology, 41(2), 292–298.

LONGHURST, A R. A survey of the fish resources
1965 of the eastern Gulf of Guinea. *J. Cons.
ICES*, 29(2), 302–334.

LUCAS, C E. Interrelationship between
1961 aquatic organisms mediated by external
metabolities. *Oceanography*, Amer. Ass.
Adv. Science Publ., 1967, 499–517.

LUMB, F H. A simple method of estimating
1963 wave height and direction over the
North Atlantic. *Mar. Obs.*, Jan. 1963,
23–29.

MAGNUSSON, J J, BRANDT, S B and STEWART, D
1979   J. Habitat preference and fishery oceanography. Lab. of Limnol. Univ. of Wisconsin-Madison, MS report.

MANKOWSKI, W. The influence of thermal
1950   conditions on the spawning of fish. *Bull. Inst. Pêches Marit.*, Gdynia, 5, 65–70.

MARR, J C. The causes of major variations in
1959   the catches of Pacific sardine (*Sardinops caerulea*, Giraud). Scient. Meet. Biol. Sardines, Rome, 1959.

MATHEWS, J P. Synopses on the biology of the
1959   South-West African pilchard (*Sardinops ocellata*, Pappe). Scient. Meet. Biol. Sardines, Rome, 1959.

McCRACKEN, F D. Distribution of haddock
1965   off the eastern Canadian mainland in relation to season, depth and bottom temperature. *ICNAF Spec. Publ. 6*, 113–129.

McKENZIE, R A. The relation of the cod to
1934   water temperatures. *Can. Fisherm.*, 21, 11–14.

McKENZIE, R A. Cod and water temperatures.
1936   *Progr. Rep. Atlantic biol. Sta.*, 17, 11–12.

MEUWIS, A L and HEUSS, M J. Temperature
1957   dependence of breathing rate in carp. *Biol. Bull.*, Woods Hole, 112, 97–107. *Water Poll. Abstr.*, 30, 7.

MEYER, P F and KALLE, K. Die biologische
1950   Umstimmung in der Ostsee in den letzten Jahrzehnten, eine Folge hydrographischer Wasserumschich-tunger? *Arch. Fischereiwsiss.* 2 (1/2), 1–19.

MIDTTUN, L. The relation between tempera-
1965   ture conditions and fish distribution in the southern Barents Sea. *Spec. Publ. ICNAF*, 6:213–9.

MITTELSTAEDT, E. Synoptische Ozeanographie
1969   in der Nordsee. *Ber. Dtsch. Wiss. Komm. Meeresforsch.*, 20(1):1–20.

MOHR, H. On the behaviour of fishes in
1960   relation to fishing gear. *Protok. Fischereitech.*, 6(29).

MUKHIN, A I and PONOMARENKO, V P. Long-
1968   term fishery prognosis of bottom fish in the Barents Sea. *Mater. Rybokhoz. Issled. Sev. Bass.*, 12:5–7 (in Russian).

MURPHY, G I. Effect of water clarity on
1959   albacore catches. *Limnol. Oceanogr.*, 4(1), 86–93.

MURPHY, G I. Population biology of the

MAGNUSSON, J J, BRANDT, S B and STEWART, D
1966   Pacific sardine (*Sardinops caerulea*). *Proc. Calif. Acced. Sciences, IV ser.*, 34(1), 1–84.

MYBERGET, S. Distribution of mackerel eggs
1965   and larvae in the Skagerrak. *Rpt. Norw. Fish. and Mar. Invest.*, 13 (8), 20–28.

NAKAI, Z. Fluctuations in abundance and
1959   availability of sardine populations caused by abiotic factors. Scient. Meet. Biol. Sardines, Rome, 1959.

NAKAI, Z, HOTTORI, S, HONJO, K and
1968   HAYASI, S. Fluctuations in the fish populations related to the environmental changes. Advances in Fisheries Oceanography. *Jap. Soc. of Fish. Oceanogr.*, 2:23. Also: *J. Coll. Mar. Sci. Technol.*, Tokai Univ., 2:115–130.

NAMIAS, J. Large-scale air-sea interactions
1963   over the North Pacific for summer 1962, through the subsequent winter. *J. Geophys. Res.*, 68(22), 6171–6186.

NELLER, W. Neue Untersuchungen über den
1965   "Schleihering", eine lokale Brackwasserform von *Clupea harengus*. *Ber. Dtsch. Wiss. Komm. Meeresforsch.*, 28(2), 162–193.

NIKOLAJEV, I I. Some factors determining
1958   fluctuations in the abundance of Baltic herring and Atlanto-Scandian herring. (In Russian with English Abstract.) *Trans. Inst. Mar. Fish. Oceanogr. USSR*, 31, 154–177.

NOMURA, M. Some knowledge on behaviour of
1958   fish schools. *Symp. Pap. IPFC*, 23, 2 pp.

NORTHCOTE, T G. Effect of photoperiodism on
1958   response of juvenile trout to water currents. *Nature*, 181, 1283–1284.

NOVIKOV, N P. Results of marking Pacific
1970   Halibut in the Bering Sea. *Izvestia TINRO Vol. 74:* 328–329.

OKONSKI, S and KONKOL, H. Daily migration
1957   of herring schools and their reaction to fishing gear. *Prace Morsk Inst. Ryback. Gdynia*, 9, 549–563.

ORTON, J H. Some inter-relations between
1937   bivalve spatfall, hydrography and fisheries. *Nature*, 140, 505.

ØSTVEDT, O J. The migration of Norwegian
1965   herring to Icelandic waters and the environmental conditions in May-June 1961-1964. *Rpt. Norw. Fish. and Mar. Invest.*, 13(8), 29–47.

ØSTVEDT, O J. Environmental data and fore-

182

1971 casting for herring and cod fisheries in Norway. *In:* The ocean world. M Uda, Ed. Proc. Jt. Oceanogr. Ass. IAPSO/ IABO/CMG/SCOR, 13-25 September, 1970, Tokyo, Japan Society for the Promotion of Science, pp. 157-8 (Abstr.)

PALOHEIMO, J E and DICKIE, L M. Production
1973 and food supply. *In:* Marine Food chains (Steele, J H, Ed.), Oliver and Boyd, Edinburgh, pp 499–527.

PARRISH, B B and CRAIG, R E. Recent changes
1957 in the North Sea herring fisheries. *Rapp. ICES*, 143(1), 12–21.

PARRISH, B B, BLAXTER, J H S and HALL, W B.
1964 Diurnal variations in size and composition of trawl catches. *Rapp. P.-v. Réun. ICES.*, 155:27–34.

PATULLO, J G and COCHRANE, J D. Monthly
1951 thermal conditions charts for the North Pacific Ocean. MS. Rept. (Scripps Inst. Oceanog.).

PEKTAS, N. Uskumrularin muhtemel
1954 muhaceret sebepleri ve bu kenuda muhit suhunetinin rolü. (The probable cause of migration of Black Sea mackerel and the part played by temperature). (In Turkish.) *Hidrobiologi Mecmuasi, Istanbul, Ser. A.2*, 113–121.

PERTSEVA, T A. Spawning, eggs and fry of fish
1939 in Motovsky Bay. Trans. *Inst. Mar. Fish. Oceanogr. USSR*, 4, 412–470.

PETERSEN, C G J and JENSEN, P B. Valuation of
1911 the sea I. Animal life of the sea bottom, its food and quantity. *Rep. Danish biol., Sta. 20*, 3–79.

PONOMARENKO, V P. Hydrological conditions
1968 and catches of bottom fish in the Southern Barents Sea. *Mater. Kybokhoz. Issled. Sev. Bass.*, 12:8–12 (in Russian).

POSTUMA, K H. The vertical migration of
1957 feeding herring in relation to light and the vertical temperature gradient. *ICES, CM. 1957, Herring Committee* (Mimeo).

POULSEN, E M. On the fluctuations in the size
1944 of the stock of cod in the waters within the Skaw during recent years. *Rep. Dan. biol. Sta.*, 46, 1–36.

RADAKOV, D V and SOLOVIEV, B S. Pervyi opit
1959 primienienia podvodnoi lodki dlia nabliudenii za poviedieniem seldi. *Rybn. Khoz.*, 35(7), 16–21.

RADOVICH, J. Some causes of fluctuations in

1959 catches of the Pacific sardine (*Sardinops caerulea*). Scient. Meet. Biol. Sardines, Rome, 1959.

RADOVICH, J. Relationships of some marine
1961 organisms of the northeast Pacific to water temperatures particularly during 1957 through 1969. *Calif. Dept. Fish and Game, Fish. Bull. 112*, 62 pp.

RADOVICH, J. Effects of water temperature on
1963 the distribution of some scombrid fishes along the Pacific coast of North America. *FAO Fish. Rep.*, (6) 3:1459–75.

RAE, K M. A relationship between wind,
1957 plankton distribution and haddock brood strength. *Bull. Mar. Ecol.*, 4(38), 347–369.

RASMUSSEN, B. Norwegian research report for
1955 1954. Note on the composition of the catch by Norwegian longliners off West Greenland. *Annual Proc. ICNAF*, 5, 43–49.

REDFIELD, A C. The history of a population of
1939 *Limacina retroversa* during its drift across the Gulf of Maine. *Biol. Bull., Mar. Biol. Lab., Woods Hole, Mass.*, 86, 26–47.

REDFIELD, A C. The effect of the circulation of
1941 water on the distribution of the Calanoid community in the Gulf of Maine. *Biol. Bull., Mar. Biol. Lab., Woods Hole, Mass.*, 80, 86–110.

REID, J L. Oceanic environments of the genus
1967 *Engraulis* around the world. *Cal. Coop. Oc. Fish. Invest., Rpt. 11*, 29–33.

RICHARDSON, J D. Some reactions of pelagic
1952 fish to light as recorded by echo-sounding. *Fish. Invest., Lond. Ser. 2*, 18 (1).

ROBINS, R C. Effects of storms on the shallow-
1957 water fish fauna of southern Florida with new records of fishes from Florida. *Bull. Mar. Sci. Gulf Caribb.*, 7(3), 266–275.

ROBINSON, M K. Summary of computer-
1966 analyzed temperature data for the Pacific and Atlantic Oceans. *Proc. Third US Navy Sympt. Milit. Oceanog.*, 243–262.

ROBINSON, M K and BAUER, R A. Atlas of
1976 North Pacific Ocean monthly mean temperatures and mean salinities of the surface layer. Wash. DC. US Naval Oceanogr. Office.

RODEWALD, M. Klima und Wetter der
1955 Fischereigebeite West- und
Südgrönland. (Climate and weather of
the fishery areas of west and south
Greenland.) *Amtl. Veröff,
Seewetteramtes*, Hamburg, 98 pp.

RODEWALD, M. Seelachs-Wanderungen unter
1960a dem Gesichtspunkt der Anderung der
Grosswetterlage. *Hansa*, 97 (19/20),
1017–1018.

RODEWALD, M. Bestandsschwankungen des
1960b Rotbarsches vor Südlabrador und die
atmosphärische Zirkulation in
Nordwest-Atlantik. *Hansa*, 97 (6/7),
365–367.

RODEWALD, M. Die jüngsten Wasser-
1960c temperatur- und Fischpendelungen in
der Barentssee als Folge von
Schwankungen der Luftzirkulation.
Hansa, 97 (8/9), 481–482.

RODEWALD, M. Die extreme 1958 und 1960 des
1960d nordatlantischen Rotbarschjahrgangs
atmosphärisch ausgelöst ? *Hansa*, 97 (18)
933–934.

RODEWALD, M. Beiträge zur Klimaschwankung
1964 im Meere. 13, Die Asymmetrie im
zeitlichen Verlauf der Wassertem-
peratur-Anomalien in Puerto Chicama
(Peru), *Dtsch. Hydrogr. Zeitschr.*, 17
(3), 105–114.

ROGALLA, E H und SAHRHAGE, D. Hering-
1960 vorkommen und Wassertemperatur.
Inform. Fischw., 7 (5/6), 135–138.

ROSA, H and LAEVASTU, T. Characteristics of
1961 the natural regions of the oceans where
sardine species occur. Scient. Meeting
Biol. Sardines, Rome, 1959.

ROUNSEFELL, G A. The existence and cause of
1930 dominant year-classes in the Alaska
herring. *Contr. Mar. Biol.*, Stanford
Univ., 260–270.

RUNNSTRÖM, S. Uber die Thermopathie der
1927 Fortpflanzung und Entwicklung mariner
Tiere in Beziehung zu ihrer geograph-
ischen Verbreitung. *Bergens Mus.
Aarb.*, 2, 1–67.

RUSSELL, F A. The vertical distribution of
1928 marine macroplankton. VIII. Further
observations on the diurnal behaviour
of the pelagic young of teleostean fishes
in the Plymouth area. *J. Mar. Biol.
Ass.* UK, 15, 829–850.

SAGALOVSKII, N V (Ed.). The influence of the
1958 North Atlantic on the development of

synoptic processes. *Gidromet.
Izdatelstvo*, Moscow (Engl. transl.)

SCATTERGOOD, L W, SINDERMANN, C J and
1959 SKUD, B E. Spawning of North American herring. *Trans. Amer. Fish. Soc.*,
88, 164–168.

SCHÄRFE, J. Fischwanderungen im Grossen
1951 Plöner See während einer Tages-
periode dargestellt an Echogrammen.
Arch. Fischereiwiss., 3 (3/4), 135–146.

SCHÄRFE, J. Uber das Verhalten von Fischen
1952 gegenüber künstlichem Licht.
Fischwirtsch, 4(9), 161–162.

SCHÄRFE, J. Uber die Verwendung künstlichen
1953 Lichtes in der Fischerei. *Protok.
Fischereitech*, 8(15), 2–29.

SCHÄRFE, J. Report on one-boat midwater-
1959 trawling experiments in the North Sea
in December 1958 and February-
March 1959. *FAO/59/11/9452*.

SCHMIDT, J. Racial investigations X. The
1931 Atlantic cod (*Gadus callarias* L.) and
local races of the same. *Comptes Rend,
Lab.* Carlsberg, 18(6), 1–72.

SCHMIDT, U. Beiträge zur Biologie des
Köhlers (*Gadus virens*) in den
isländischen Gewässern. *Ber. Dtsch.
Wiss. Komm. Meeresforsch.*, 14, 46–85.

SCHOTT, G. Geographie des Indischen und
1931, Stillen Ozeans. Geographie des
1942. Atlantischen Ozeans. Hamburg,
Boysen.

SECKEL, G R and WALDRON, K D. Oceano-
1960 graphy and the Hawaiian skipjack
fishery. *Pac. Fish.*, February 1960.

SECKEL, G R. Hawaiian-caught skipjack tuna
1972 and their physical environment. *Fish.
Bull. NMFS/NOAA*, 70(3):763–87.

SETTE, O E. Biology of the Atlantic mackerel
1950 (*Scomber scombrus*) of North America.
Part II. Migrations and habits. *US Fish
and Wildl. Serv., Fish Bull.*, 49, 251–258.

SETTE, O E. Problems in fish population
1961 fluctuations. *Rep. Calif. Coop. Fish.
Invest.*, 8, 21–24.

SHAPIRO, S. The Japanese long-line fishing for
1950 tunas. *Comm. Fish. Rev.*, 12(4), 1–26.

SHAW, E. Minimal light intensity and the
1961 dispersal of schooling fish. *Bull. Inst.
Oceanogr.*, Monaco, 58 (12/13), 8 pp.

SIMONSEN RADIO. Fish-finding with sonar.
1964 Oslo, Simonsen Radio (Publ), 96 pp.

SIMPSON, A. C. Some observations on the
1953 mortality of fish and the distribution of

plankton in the southern North Sea during the cold winter 1946-1947. *J. Cons. ICES*, 19, 150–177.

SJÖBLOM, VEIKKO. Wanderungen des
1961 Strömlings (*Clupea harengus* L.) in einigen Schären- und Hochseegebieten der nördlichen Ostsee. *Ann. Zool. Soc. 'Vanamo'*, 23 (1), 1–193.

STEELE, J H. Stability of plankton ecosystems.
1974 *In* Ecological Stability., Usher, M B and Williamson, M H (Eds.): 179-191, Chapman & Hall, London.

STEVENSON, W H and PASTULA, E J, Jr.
1971 Observations on remote sensing in fisheries. *Commer. Fish. Rev.*, 33(9): 9–21.

SULLIVAN, C M and FISHER, R C. Seasonal
1953 fluctuations in the selected temperature of speckled trout (*Salvelinus fontinalis*). *J. Fish. Res. Bd. Canada 10*, 187–195.

SULLIVAN, C M. Temperature reception and
1954 responses in fish. *J. Fish. Res. Bd. Canada, 11* (2), 153–170. *Water Poll. Abstr.* 30, 7.

SUTCLIFFE, W H, Jr, DRINKWATER, K and
1977 MUIR, B S. Correlations of fish catch and environmental factors in the Gulf of Maine. *J. Fish. Res. Board Canada*, 34:19–30.

SVERDRUP, H B and MUNK, W H. Wind, sea
1947 and swell: theory of relations for forecasting. *US Navy Hydrographic Office Publ.*, 601, 44 pp.

TAKAYAMA, S. Sauri lift-net fishing with light.
1949 UN ECOSOC, E/Conf. 7/SEC/W, 184, 25 April, 1949. *Wildlife*, 5 (5), (Mimeo).

TÄNING, A V. North-Western area. Intro-
1951 duction. *Ann. biol. ICES*, 7 (8).

TÄNING, A V. (Articles on long-term
1953 temperature changes and fisheries). *Annual Proc. ICNAF* 3, 1952–53.

TAYLOR, C C. Cod growth and temperature.
1958 *J. Cons. ICES*, 23, 366–370.

TEMPLEMAN, W. Anomalies of sea temperature
1964 at Station 27 off Cape Spear and of air temperature at Torbay – St. John's. ICNAF Environmental Symp., Rome Jan. 1964 (ms).

TEMPLEMAN, W and HODDER, V M. Distribu-
1965 tion of haddock on the Grand Bank in relation to season, depth and temperature. *ICNAF Spec. Publ.*, 6, 171–187.

TERADA, K. Utilization of sea surface
1959 temperature data for fisheries and the special forecast for fishing operations in Japan. Paper, presented at CMM III, Utrecht (Mimeo).

TESTER, A L. Herring, the tide and the moon.
1938 *Prog. Rep. Biol. Stan. Nanaimo and Prince Rupert*, 38, 10–14.

THOMPSON, H. A biological and economic
1943 study of cod (*Gadus callarias* L.) in the Newfoundland areas. *Res. Bull. Div. Fish. Res. Newfoundland*, 14.

TIEWS, K. The role of whiting as an undesir-
1963 able guest in German coastal waters. *Veröff. Inst. für Küsten- und Binnenfischerei*, Hamburg, 27, 9–19.

TOLSTOY, I and CLAY, C S. Ocean acoustics.
1967 New York, McGraw-Hill., 293 pp.

TOMCZAK, G H. Environmental analysis in
1977 marine fisheries research – fisheries environmental services. *FAO Fish. Techn. Pap.* 170:141 pp.

TROUT, G C. The Bear Island cod: migrations
1957 and movements. *Fish. Invest.* London, Ser. 2(21), (6), 51 pp.

TSUJITA, T. The fisheries oceanography of the
1957 East China Sea and Tsuchima Strait. 1. The oceanographic structure and the ecological character of the fishing grounds. *Bull. Seikai Reg. Fish. Res. Lab.*, 13, 1–47.

TUCKER, D G. Sonar in fisheries. London,
1966 Fishing News (Books) Ltd.

TULLY, J P and GIOVANDO, L F. Seasonal
1963 temperature structure in the eastern subarctic Pacific Ocean. *Royal Soc. Canada, Spec. Publ.*, 5, pp 10–36.

TYURNIN, B V. The spawning range of
1973 Okhotsk herring. *Izv. Tikhookean. Nauchno-Issled. Inst. Rybn. Khoz. Okeanogr.*, 86:12–21 (in Russian). Issued also as Transl. *Ser. Fish. Mar. Serv. Can.*, (3247):18 p. (1974).

UDA, M. Fishing centre of 'sanma' (*Cololabis
1936 saira*), correlated with the head of Oyashio cold current. (In Japanese with English summary) *Bull. Jap. Soc. Scient. Fish.*, 5, 236–238.

UDA, M. On the relation between the variation
1952 of the important fisheries conditions and the oceanographical conditions in the adjacent waters of Japan. 1. *J. Tokyo Univ. Fish.*, 38, 363–389.

UDA, M. Water mass boundaries – 'Siome'.

1959a Frontal theory in oceanography. *Fish. Res. Bd. Canada, MS Rpt., Series 51,* 10–20.

UDA, M. Intrusion and isolated water masses.
1959b *Fish. Res. Bd. Canada, MS Rpt. 51,* 21–27.

UDA, M. The fisheries of Japan. *Fish. Res. Bd.*
1959c *Canada, Nanaimo Biol. St.* (Mimeo), 96 pp.

UDA, M. The fluctuation of the sardine fishery
1959d in the oriental waters. Scient. Meet. Biol. Sardines, Rome, 1959.

UDA, M. Fishery oceanography of the western
1974 Pacific: application of oceanographic information to forecast natural fluctuations in the abundance of certain commercially important pelagic fish stocks. *Proc. IPFS,* 15(3):56–65.

UDA, M and HONDA, K. The catch of keddle
1934 nets in two fishing grounds on the coast of Nagasaki Prefecture and Izu Peninsula. *Bull. Jap. Soc. Scient. Fish,* 2, 253–271.

UDA, M and OKAMOTO, G. Effect of oceano-
1936 graphic conditions on 'Iwashi' (sardine) fishing in the Japan Sea. *Journ. Imp. Fish. Exper. Sta.,* 7, 19–49.

UDA, M and WATANABE, N. Autumn fishing of
1938 skipper and bonito influenced by the rapid hydrographic change after the passage of cyclones. *Bull. Jap. Soc. Scient. Fish.,* 6, 240–242.

URICK, R J. Principles of underwater sound.
1975 McGraw-Hill, New York. 384 pp.

URIK, R J. Principles of underwater sound for
1967 engineers. NY, McGraw-Hill, 342 pp.

VESTNES, G. Classification of sonar echos.
1964 *In:* Fish-finding with sonar. Oslo, Simrad, 27–41.

VESTNES, G. Tactical applications. *In:* Fish-
1964 finding with sonar. Oslo, Simrad, 42–66.

VESTNES, G and JAKOBSON, J. Search methods.
1964 *In:* Fish-finding with sonar. Oslo, Simrad, 10–26.

WALDEN, H and SCHUBERT, K. Untersuchungen
1965 über die Beziehungen zwischen Wind und Herings-Fangertrag in der Nordsee. *Ber. Dtsch Wiss. Komm. Meeresforsch,* 28 (2), 194–221.

WALFORD, L A. Effect of current on distri-
1938 bution and survival of the egg and larvae of the haddock (*Melanogrammus aeglefinus*) on Georges Bank. *Fish. Bull. US Bureau of Fish,* 49 (29), 73 pp.

WISE, J P. Cod and hydrography: a review.
1958 *US Fish Wildl. Serv., Scient. Rep.,* 245, 1–16.

WISE, J P. Migration of cod (*Gadus morhua*
1959 L.). *Prepr. Int. Oceanogr. Congr., New York,* 358–359.

WITTING, R. Zur Kenntnis den vom Winde
1909 Erzeugtem Oberflächenstromes. *Ann. Hydrogr. Marit. Met.,* 73:193 pp.

WOLFF, P M. Numerical synoptic analysis of
1967 sea surface temperature. *Int. J. Oceanol. and Limnol.,* 2(7), 277–290.

WOODHEAD, A D. The migrations of Arctic
1959 cod. *New Scient.,* 6(154), 796–798.

WOODHEAD, P M J. Effects of light upon
1965 behaviour and distribution of demersal fishes of the North Atlantic. *ICNAF Spec. Publ.,* 6, 267–278.

WOODHEAD, P M J and WOODHEAD, A D.
1955 Reactions of herring larvae to light: a mechanism of vertical migration. *Nature,* 176 (4477), 349–350.

YAMANAKA, H. Relation between the fishing
1969 grounds of tuna and the equatorial current system. *Bull. Jap. Soc. Fish. Oceanogr.,* Spec. No: 227–30.

YAMANAKA, I. Oceanography in tuna research.
1978 *Rapp. P.-v. Réun. ICES,* 173:203–211.

ZUSSER, S G. A contribution to the study of fish
1958 behaviour. Paper presented to the IPFC Symp. on Fish Behaviour.

13

Explanation of terms

ABC

Acceptable Biological Catch. Subjectively estimated amount of catch of given species from a given region. Has no objective scientific definition.

Absorption of light

When light penetrates into water, it is gradually absorbed, that is, transformed into heat. The rate of absorption, which is not the same for all wave-lengths, is measured as the difference between extinction and scattering.

Acclimatization

is the phenomenon through which an organism becomes habituated to a climate not native. The same term is used also when referring to an 'aquatic climate', that is, to environmental conditions in the sea.

Advection

or advective water movement means basically horizontal transport of waters (by currents).

Anticyclonic

as a term explaining the direction of rotation, means in the same direction as that of the winds around a high pressure area in the atmosphere, or an anticyclone. In the northern hemisphere this is clockwise; in the southern hemisphere, anticlockwise.

Bar (millibar)

Measure of pressure in the atmosphere and in the ocean. 1mb (millibar)=0·750mm Hg.

Baroclinic

Condition in the ocean and in the atmosphere where constant pressure surfaces intersect constant density surfaces.

Barotrophic

Assumed (simplified) condition in the atmosphere where constant pressure and constant density surfaces are identical.

187

Bathythermograph	BT, is a mechanical instrument used for measuring and indicating temperature and depth on a smoked glass slide as the BT instrument is towed through the water. It gives a continuous trace on the slide of the water temperature against the depth.
Benthos	Animals and plants which live on or in the bottom, and also those whose life is in some way closely connected with the bottom.
Biogeography	The study of the geographical distribution of animals and plants. Marine biogeography includes description of the distribution of marine animals and plants and analyses of factors which determine the distribution and abundance of a given species.
Biomass	is the wet weight or mass of living organisms.
Biota	refers to all fauna and flora.
Biotic factors	Factors of a biological nature, such as availability of food, competition between species, predator and prey relationships, *etc*, which, besides the purely environmental factors, also affect the distribution and abundance of a given species of plant or animal.
Carrying capacity	The amount of biota (*eg* finfish) which a defined ocean area can sustain.
Catabolism	Destructive metabolism, destruction of living matter.
Chlorinity	is defined as the 'total amount of chlorine, bromine and iodine in grams contained in one kilogram of sea water, assuming that the bromine and the iodine had been replaced by chlorine'. Ordinarily the 'chlorides' are determined by titration with silver nitrate. Since the quantitative relations of major elements in sea water are remarkably constant, the salt content, the so-called 'salinity' of sea water, can be computed from chlorinity.
Cohort	age-class of a species (fish).
Cold front	is an air mass boundary between cold and warm air in which the cold mass is pushing back the warm air.
Community	of animals, which live in the same locality under the influence of similar environmental factors and affect the existence of each other through their activities. The

communities are usually named by the dominant species in them.

Contranatant Swimming or migrating against the current (usually adult fish, migrating towards the spawning areas).

Convection in oceanography means vertical water movements while the term 'advection' is saved for horizontal movements. The convection may be either thermohaline or dynamic. The thermohaline convection is observed when sea water, because of its decreasing temperature or increasing salinity, has become heavier than the water underneath it and a disturbed vertical equilibrium results. Dynamic convection could be called 'forced vertical mixing'.

Convergence is observed in the current field when waters flow (=verge) together, observed sometimes as a 'point' but normally as a line. Most convergence lines are one-sided, in which case the water mass converging to the line dives under the other water mass. Convergence lines represent in most cases boundaries between different water types.

Cyclogenesis Formation of cyclones (anticlockwise circulation cells) in the atmosphere.

Cyclone, extra-tropical is a term used commonly for the low-pressure cells of the middle and higher latitudes. These extra-tropical cyclones normally develop from wavelike outbreaks along frontal surfaces and consist of a warm front and of a cold front. The cold front moves faster than the warm front and therefore finally overtakes it, lifting the warm air mass off the ground. This stage of development of the cyclone is called occlusion.

Cyclonic as a term explaining the direction of rotation, means in the same direction as that of the winds of a cyclone. In the northern hemisphere cyclonic is anticlockwise, *ie* against the sun.

Cyclindrical spreading Spreading of sound waves in a sound channel or between the surface and bottom.

Decibel (dB) Measure for the loudness of sound (logarithm of power of sound).

Demersal means 'living near the bottom' (from the Latin *demersus*, meaning 'plunged under').

189

Denatant   Swimming, drifting, or migrating with the current (usually eggs and larvae).

Detritus   Sometimes also called 'tripton', means dead organic debris.

Divergence   The opposite to convergence, refers to waters flowing apart or diverging. Divergence is observed either at a 'point' or from a line or zone. Obviously the diverging waters must be replaced by water upwelling from deeper layers of the ocean.

Environmental factors   consist of physical and chemical conditions in the environment (water), such as temperature, salinity, light conditions, current velocity, *etc*. The 'biotic factors' are usually not included in the environmental ones.

Epifauna   Benthic organisms living on the surface of the sea bottom.

Euryhaline   Organisms are said to be euryhaline if they can tolerate a wide range of salinity.

Exploitable biomass   The biomass of fishable size fish in a given stock.

Extinction of light   When light penetrates into water, it gradually dies out; it is extinguished because of absorption and scattering. The rate of extinction is not the same for all the wave-lengths.

Fish stock   Fish of a given species, living in a given area and using the same spawning and feeding areas.

Fisheries hydrography   Nowadays frequently called also 'fisheries oceanography' (see below).

Fisheries oceanography   is the study and application of oceanography, maritime meteorology and aquatic ecology to certain practical problems in fisheries. These practical problems are related to the productivity of the oceans or to the fisheries' resources, to the behaviour of various specimens, to the availability of fish and other fishable marine animals to the fisheries, and to the effects of oceanographic and meteorological conditions on the conduct of a fishery.

Gradient   in the general sense, means the rate of change of a property in a certain direction.

Grid in numerical computation is a net of squares (and/or quadrangles) drawn over an area where computation is made at each grid intersection (see also mesh).

Holistic ecosystem (simulation) Numerical simulation of (marine) ecosystem which contains all biota, environment and all essential processes within this ecosystem.

Homothermal from the Greek word *homos*, meaning identical to something, and *therme*, heat. It refers to sea water of like or uniform temperature. It is a common practice to call a water mass of uniform temperature 'homothermal' and to reserve the word 'isothermal' (like temperature) to surfaces or lines along which the temperatures are the same.

Hydroclime Climatology of the oceans (surface and below it).

Hydrodynamical numerical (HN) model Numerical model for computation of sea level changes, currents, and any desired distribution of properties. These models use basic (primitive) equations of motion and equations of state.

Hydrography is the science and practice of surveying and charting the oceans and the seas. It excludes biological problems.

Hydropsis is a term given recently to that part of oceanography concerned with continuous observations and data collecting and the reporting of oceanographic phenomena on a regular and prompt basis with the aim of supplying those that use the seas, particularly in commercial fishing, current information. Similarly, weather reporting is based on synoptic meteorology. Synoptic because of the necessity of simultaneous, or synoptic, observations. The routine of synoptic meteorology is called 'synopsis'. Therefore, the routine of synoptic oceanography has been given the name 'hydropsis'.

IGOSS Integrated Global Ocean Station System; an acronym used to designate ocean weather ships and measuring buoys.

Infauna Benthic organisms which are partially or fully buried in the sediment.

Isobar	refers to the Greek word *isos*, meaning equal, and to the pressure unit 'bar' (or the better known millibar). Thus isobar or 'isobaric' refers to equal pressures.
Isohaline	refers to waters or sea areas of equal or like salinities. Generally, on a chart, isohaline lines or curves are drawn connecting points of like salinities.
Isotherm	refers to waters, areas, air masses, *etc* of like temperature. Along an 'isothermal surface' the temperatures are alike.
Mechanoreceptor (organ)	Organs in fish (mainly along lateral line), which enable the fish to sense currents and its own movement.
Mesh (standard, small, *etc*)	Computational grid (see this word). Standard grid refers to a conventional hemispheric grid with mesh size of *ca* 390km.
Metabolism	meaning actually a 'change', refers to chemical changes taking place within living material (reactions within an organism).
Metaphysics	Deduction of results and describing of nature by deducing logical conclusions of reality and of being, based on known facts about related being and phenomena.
Meteorological front	or air mass boundary=frontal surface, is formed when two air masses of at least somewhat different temperature and moisture collide. Because of the said differences, instead of getting mixed with each other, the colder air mass squeezes itself under the warm one and lifts it. It is easy to observe in the oceans analogous water mass boundaries which could similarly be called 'oceanographic fronts'.
MLD	Mixed Layer Depth, top of the seasonal thermocline (see further thermocline).
MSY	An average over a reasonable length of time of the largest catch which can be taken continuously from a stock under current environmental conditions. It should normally be presented with a range of values around its point estimate.
Nekton	Mobile aquatic animals which have the ability (in spite of currents and waves) to actively determine their course of movement.

Nutrients	in sea water refers to elements required to support the growth of phytoplankton in the sea. These include usually phosphates, nitrates, and silicates, but sometimes also the minor elements of sea water, such as copper, manganese, cobalt, iron, *etc*, are included in this term.
Olfaction	The sense of smell or faculty of smelling.
Oligohaline	Oligohaline organisms are those marine organisms which tolerate narrow range of salinity.
Osmoregulation	Regulation of the concentration of dissolved substances in the body fluids (mainly by diffusion through semi-permeable tissues/membranes).
Oxygen minimum layer	usually observed in the tropical seas, is a subsurface layer, in which the dissolved oxygen content is very low or nil.
OY	Optimum Yield, in reality undefinable quantity, used in some fisheries management considerations.
Parameter	of a mathematical function is a quantity to which the operator may assign any arbitrary value, as distinguished from a variable, which can assume only those values that the function makes possible.
Pelagic	refers to all organisms living freely in the water masses. Analogously the mid-water trawl is called pelagic trawl.
Photosynthesis	is the phenomenon upon which the existence of all the plants is based: in the presence of carbon dioxide and water, plants are able to intercept the sun's radiation energy required by them in the manufacture of compounds of carbon. In the sea, the photosynthesis is basically limited to the euphotic zone.
Phototaxis	is a movement, stimulated by light, either toward the light, photopositively, or away from the light, photonegatively.
Pineal gland	A gland in the upper region of fish brain which can sense light.
Phytoplankton	A general term for all freely floating, mainly microscopic, plankton algae of the waters (see also plankton).
Plankton	A general term for all the organisms of the free water masses, both for plants and animals, without own

locomotion or with such a limited swimming ability that they are forced to float passively with the water movements. Plankton consists both of phytoplankton, with its minute algae, and of zooplankton (animal plankton).

Polar fronts of the ocean are called also Arctic Convergences and Antarctic Convergences. These more or less permanent boundaries separate the cold polar waters from the warmer waters of the middle latitudes.

Predator An animal living by preying on other animals.

Primitive equation (PE) model Numerical models which use basic equation of motion and equation of state.

Propagation loss Loss of sound energy during its propagation through water (mainly absorption and scattering losses).

Pycnocline or 'density discontinuity layer' is a horizontal layer with a limited thickness at which density vertically changes abruptly. Thus both the thermocline and the halocline are pycnoclines. However, rather frequently the term pycnocline is used in the sense of 'halocline'.

Rheotaxis Response of the fish to current.

Salinity The content of salts in sea water, expressed as parts per thousand (or g per kg of water) (see further chlorinity).

Scattering of light refers to that part of the light penetrating into water, which is reflected from the water molecules, particles, and gas bubbles in the sea.

Section A two-dimensional picture of distribution of properties and/or biota. Also a line of observation and/or collecting stations.

Sessile means permanently attached, not free to move about.

Simulation Quantitative (numerical) reproduction of marine ecosystem, based on all available empirical knowledge of this system.

Sonar (asdic) An electrical system which sends out a horizontal (or quasi-horizontal) beam of sound in the water (with defined frequency) and records the return of echoes of this sound.

194

Sonic Layer Depth (SLD)	Depth of the layer below the surface where sound speed is at maximum (usually same as MLD).
Sound channel (duct)	A depth range around a sound speed minimum (*eg* between surface and SLD) (see SLD).
Spherical spreading	Spreading of sound in all directions from a point source.
Stock recovery	A term signifying the increase of a heavily fished stock after decrease of fishing effort.
Subthermocline duct	Sound duct (channel) below thermocline, bounded with thermocline and another sound speed maximum below it.
Surface duct	Sound channel (duct) between surface and SLD (see SLD).
Synoptic	means simultaneous. In the daily weather service the weather observations made at various places for obvious reasons must be simultaneous. Thus 'actual meteorology' has been given the name 'synoptic meteorology'. Analogously 'actual oceanography', eventually leading to an information service on the actual phenomena in the oceans, must be based on more or less synoptic data. Therefore: synoptic oceanography or synoptic hydrography is similarly called 'hydropsis'.
Tactile sense	is the sense of touch.
Taxonomy	from the Greek words *taxis*=arrangement, and *nomos* rule, meaning the doctrine of classification of plants and animals according to their natural relationships.
Temperature inversion	Normally in the sea, at least in the uppermost layers, the temperature decreases with the depth. If at a certain depth the vertical change of temperature is inverse, increasing instead of decreasing, a 'temperature inversion' is said to be observed. The layer where this occurs is called an 'inversion layer'.
Thermocline	or 'temperature discontinuity layer' is a horizontal layer with a limited thickness at which temperature vertically changes abruptly. Since the differing temperatures make the waters above and below the thermocline different in their density, they do not mix easily with each other. Thus the thermocline separates vertically two 'water masses'.

TSD

Temperature-salinity-depth recorder, a recording instrument lowered from research vessels.

Turbidity

State of water not being clear or translucent, due to suspended fine particles in it.

Turbulence

in the sea refers to the mixing of the waters through the agitating action of numerous turbulent eddies of various sizes moving in a variety of directions. Without turbulence a continuous injection of narrow beam of a dye would move laminarily and would remain a narrow beam over very long distances regardless of its path. With turbulence the dye is rapidly mixed both horizontally and vertically with the waters neighbouring the injection site.

Upwelling

of deeper waters, rich in nutrient salts, occurs usually as the result of relatively steady winds in the tropics and subtropics blowing along the coasts and/or of the wind divergences in the equatorial region. The upwelling, though brought about purely mechanically, is biologically most significant.

Vagile

means mobile, with own locomotion. (Vagile sense is sense of movement.)

Water pocket

A water mass of limited size, frequently in the form of a pocket and having a different property (*eg* temperature) from the surrounding water mass. 'Water pockets' are often located at a meandering boundary where they usually are partly or wholly cut off from their original main body of water.

Wave spectrum

is a concept, in a sense, analogous to the well-known energy spectrum of light. In the sea normally waves with differing periods are observed simultaneously. A graphic representation where the period of the measured waves is given as abscissa with their occurrence frequencies as ordinate indicates the state of sea better than, say, the 'mean period' of all the simultaneously appearing waves.

XBT

Expendable bathythermograph; an instrument which is lowered from the ship to measure temperature changes with depth, and is not retrieved.

Zooplankton

is a general term for all the aquatic animals with such a limited swimming ability that they are forced to float passively with the movements of water.

14

Subject index

Fishing with light
Freezing and irradiation of fish
Handbook of trout and salmon diseases
Handy medical guide for seafarers
How to make and set nets
Inshore fishing: its skills, risks, rewards
Introduction to fishery by-products
The lemon sole
A living from lobsters
Marine fisheries ecosystem: its quantitative evaluation
    and management
Marine pollution and sea life
The marketing of shellfish
Mending of fishing nets
Modern deep sea trawling gear
Modern fishing gear of the world 1
Modern fishing gear of the world 2
Modern fishing gear of the world 3
More Scottish fishing craft and their work
Multilingual dictionary of fish and fish products
Navigation primer for fishermen
Netting materials for fishing gear
Pair trawling and pair seining: the technology of two boat fishing
Pelagic and semi-pelagic trawling gear
Planning of aquaculture development: an introductory guide
Power transmission and automation for ships and submersibles
Refrigeration on fishing vessels
Salmon and trout farming in Norway
Salmon fisheries of Scotland
Scallops and the diver-fisherman
Seafood fishing for amateur and professional
Seine fishing: bottom fishing with rope warps and wing trawls
Stability and trim of fishing vessels
Study of the sea
The stern trawler
Textbook of fish culture: breeding and cultivation of fish
Training fishermen at sea
Trout farming manual
Tuna: distribution and migration
Tuna fishing with pole and line